RIGHTSTART™ MATHEMATICS

by Joan A. Cotter, Ph.D.
with Kathleen Cotter Lawler

LEVEL E WORKSHEETS

Second Edition

Activities for Learning, Inc.

A special thank you to Maren Ehley and Rebecca Walsh for their work in the final preparation of these worksheets.

Printed in the United States of America

www.RightStartMath.com

For more information: info@RightStartMath.com
Supplies may be ordered from: www.RightStartMath.com

Activities for Learning, Inc.
321 Hill Street
Hazelton, ND 58544-0468
United States of America
888-775-6284 or 701-782-2000
701-782-2007 fax

ISBN 978-1-942943-12-9
May 2017

Game Log

Math games are an important part of the RightStart™ Mathematics program. Instead of flash-cards or timed tests, games supply the necessary repetition for understanding and mastery in an enjoyable setting, providing a superior educational atmosphere for the child. The lessons indicate which games to play. Additional games can be found in the *Math Card Games* book.

This Game Log provides a record of the games played and when they were played.

Date	Game Played	Players

Games help children understand, apply, and enjoy mathematics.

GAME LOG CONTINUED

Date	Game Played	Players

Games help children understand, apply, and enjoy mathematics.

Game Log continued

Date	Game Played	Players

Games help children understand, apply, and enjoy mathematics.

Game Log continued

Date	Game Played	Players

Games help children understand, apply, and enjoy mathematics.

Name: ______________________________

Date: ______________________________

1. Kaitlyn is selling tickets for a sporting event. Nine people can sit in a row in the bleachers. A group of 12 people are attending. How many full rows of nine do they need? How many people will be sitting in the row that is not full, the partial row? Write your answers in the left chart below.

2. Next Kaitlyn sells tickets to a group of 22 people. How many full rows of nine will they need? How many will sit in the partial row? Write your answers in the left chart.

3. Kaitlyn then sells tickets to a group of 28 people. How many full rows will they need? How many will sit in the partial row? Write your answers in the left chart.

4. At halftime the three groups left, returned, and sat together. What is the fewest number of rows they need now? Write your answers in the left chart.

Kaitlyn's Ticket Sales		
Number of People	Full Rows with 9 People	People in Partial Row
12		

Kurt's Ticket Sales		
Number of People	Full Rows with 9 People	People in Partial Row

5. Kurt is also selling tickets. His first group has 15 people, the second group has 23 people, and the third group has 30 people. Find the number of full rows and the number of people sitting in partial rows per group. Write your answers in the right chart above.

6. At halftime the three groups left, returned, and sat together. What is the fewest number of rows Kurt's groups need now? Write your answers in the right chart above.

7. What do you notice in Kaitlyn's chart about the number of people sitting in the partial rows before and after the groups are combined? ______________________________

____

8. What do you notice in Kurt's chart about the number of people sitting in the partial rows before and after the groups are combined? ______________________________

____

Worksheet 2, Using Check Numbers to Check Adding

Name: ______________________________

Date: ______________________________

Add these numbers. Use check numbers to check your work.

( ) ( ) ( ) 7 + 9 = ____	( ) ( ) ( ) 18 + 30 = ____	( ) ( ) ( ) 77 + 6 = ____
( ) ( ) ( ) 41 + 21 = ____	( ) ( ) ( ) 19 + 35 = ____	( ) ( ) ( ) 25 + 29 = ____
( ) ( ) ( ) 73 + 37 = ____	( ) ( ) ( ) 56 + 24 = ____	( ) ( ) ( ) 43 + 40 = ____
( ) ( ) ( ) 78 + 20 = ____	( ) ( ) ( ) 75 + 26 = ____	( ) ( ) ( ) 68 + 24 = ____
( ) ( ) ( ) 147 + 33 = ____	( ) ( ) ( ) 188 + 26 = ____	( ) ( ) ( ) 209 + 428 = ____
( ) ( ) ( ) 39 + ____ = 100	( ) ( ) ( ) 87 + ____ = 104	( ) ( ) ( ) 127 + ____ = 200

Name: ____________________________

Date: ____________________________

Warm-Up

Add these numbers. Use check numbers to check your work.

( ) ( ) ( )
14 + 23 = _____

( ) ( ) ( )
29 + 41 = _____

( ) ( ) ( )
78 + 21 = _____

Use check numbers for all additions.

Add the number on the abacus plus 2576.

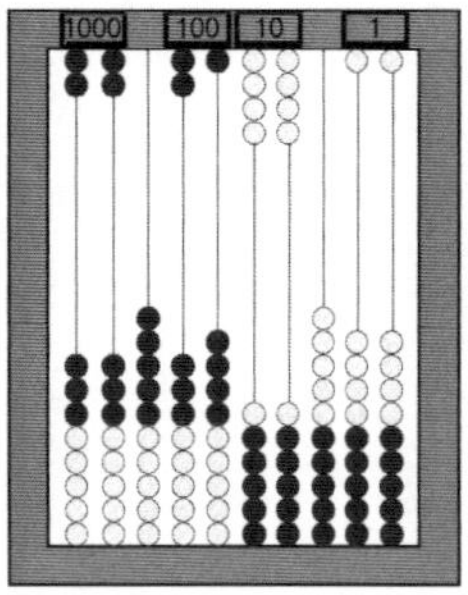


```
   4382 ( )
 + 2576 ( )
```

Add the number on the abacus plus 4078.

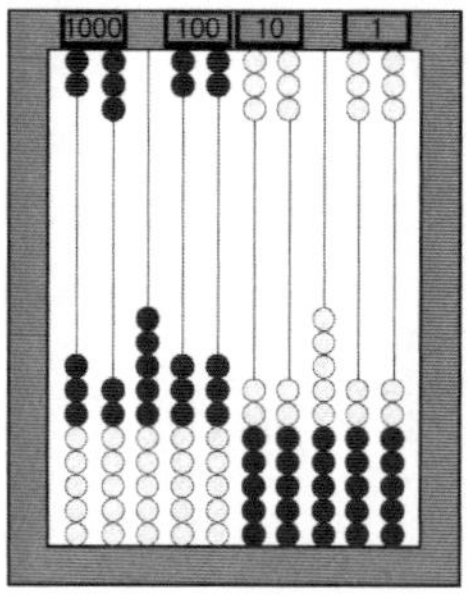

Add the number on the abacus plus 1909.

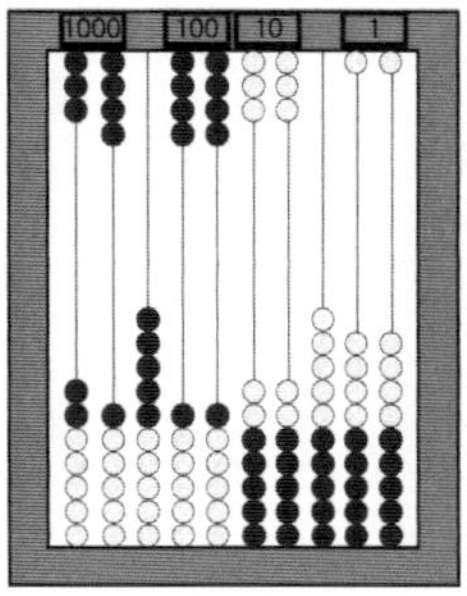

Add the number on the abacus plus 4795.

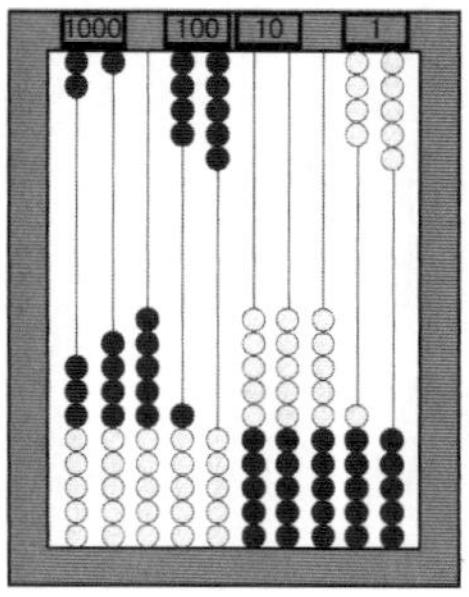


```
   4382 ( )
 + 2576 ( )
```

```
   4567 ( )
 + 7859 ( )
```

```
   5997 ( )
 + 3997 ( )
```

Name: ______________________________

Date: ______________________________

Warm-Up

Add these numbers. Use check numbers to check your work.

4382 ( )	2374 ( )	2988 ( )
+ 5948 ( )	+ 5958 ( )	+ 5247 ( )

INFORMATION: When writing numbers with digits, put a comma between number periods.

Underline the period words in the following numbers. Then compose the number with place-value cards and write it in the rectangles. Write the number on the line using commas.

five billion four hundred thirty-one million seven hundred two thousand eight hundred sixty-nine

☐ ☐☐☐ ☐☐☐ ☐☐☐ ______________________________

thirty-two billion nine hundred seventeen million five hundred eighty-four thousand six hundred

☐☐ ☐☐☐ ☐☐☐ ☐☐☐ ______________________________

sixty billion two hundred ninety-eight million eight hundred thousand one hundred thirty-four

☐☐ ☐☐☐ ☐☐☐ ☐☐☐ ______________________________

eleven billion twenty-nine million eight hundred fifty-five thousand seventy-three

☐☐ ☐☐☐ ☐☐☐ ☐☐☐ ______________________________

Write these number in words:

305,162,000 ____

____

200,064,014 ____

____

Name: ____________________

Date: ____________________

Warm-Up

Fill in the tables.

	Quotient (answer)	Remainder
49 ÷ 7		
50 ÷ 7		
51 ÷ 7		
52 ÷ 7		

	Quotient (answer)	Remainder
27 ÷ 6		
28 ÷ 6		
29 ÷ 6		
30 ÷ 6		

Round the following to the hundreds place and put the number into the correct columns.

12,323	12,470	12,424	12,567	12,400
12,491	12,399	12,603	12,599	12,250
12,290	12,461	12,546	12,287	12,356

12,300	12,400	12,500	12,600
12,323			

The earth's population on October 23, 2015 was estimated to be 7,375,870,314.

What is the earth's population to the nearest million? ____________________

What is the earth's population to the nearest billion? ____________________

Round the populations to the nearest million.

Australia	23,630,169	_______ million
Canada	35,524,732	_______ million
China	1,404,643,511	_______ million
Ethiopia	99,756,649	_______ million
United Kingdom	64,596,800	_______ million
United States	325,937,137	_______ million
World	7,375,870,314	_______ million

Write the countries in order from least population to greatest.

Name: ______________________________

Date: ______________________________

Write only the answers.

____

____

____

Write the answers.

318 + 87 = ______

43 + ______ = 200

(6 × 3) + (6 × 4) = ______

Add. Use check numbers.

```
  4928 ( )
+ 2566 ( )
```

Write these numbers using digits and commas.

39 million 24 thousand 718 ______________________

1 billion 319 thousand 679 ______________________

623 million 381 ______________________

Fill in the table.

	Round to nearest hundred.	Round to nearest thousand.	Round to nearest million.
6,083,140			
18,945,562			
65,709,697			
2,376,450,712			

Write >, <, or = on the lines.

6 million _____ 3,000,000 × 2

10,000 × 10 _____ 1,000,000

29 × 1000 _____ 29,000

3 hundred 6 _____ 3006

40 × 9 _____ 6 × 60

30 × 20 _____ 30 + 2

Fill in the table.

	Quotient (answer)	Remainder
42 ÷ 2		
24 ÷ 7		
73 ÷ 9		
60 ÷ 9		
24 ÷ 10		
102 ÷ 10		

Name: ____________________________

Date: ____________________________

Write only the answers.

____

____

____

Write the answers.

453 + 87 = ______

64 + ______ = 200

(7 × 5) + (7 × 2)= ______

Add. Use check numbers.

```
  3786  ( )
+ 5492  ( )
```

Write these numbers using digits and commas.

73 million 853 thousand 37 ____________________

18 million 46 thousand 679 ____________________

2 billion 840 million 7 thousand 832 ____________________

Fill in the table.

	Round to nearest hundred.	Round to nearest thousand.	Round to nearest million.
5,935,899			
56,057,619			
89,767,846			
3,587,123,777			

Write >, <, or = on the lines.

6 million _____ 2,000,000 × 3

100,000 _____ 10,000 × 100

74 × 1000 _____ 74,000

5009 _____ 5 hundred 9

30 × 8 _____ 6 × 40

500 + 20 _____ 500 × 2

Fill in the table.

	Quotient (answer)	Remainder
28 ÷ 2		
37 ÷ 5		
73 ÷ 10		
98 ÷ 10		
80 ÷ 9		
70 ÷ 9		

Name: ______________________

Date: ______________________

Add using shortcuts.	Add starting from the left.		
27 + 69 = _______	191 + 323	10,298 + 4,375	6537 + 5948
354 + 78 = _______			
898 + 326 = _______			
Subtract using shortcuts.	**Subtract using shortcuts.**		
67 – 18 = _______	787 – 768	238 – 209	845 – 547
46 – 37 = _______			
234 – 135 = _______			

Write a 3-digit number with consecutive digits. Then reverse the digits. For example, the first number could be 678, then the second number would be 876. Subtract the smaller number from the larger number. Do this at least three times and note the pattern.

Does this work for 4-digit numbers? For 5-digit numbers?

Write any 3-digit number. Then reverse the digits. Find the difference. Then reverse the digits of the difference and add it to original difference. For example, if the first number is 149:

941 – 149 = 792; 792 + 297 = _______

Try this with these numbers, 412, 863, and 356. Note the pattern.

Name: ______________________

Date: ______________________

Warm-Up

Add using shortcuts.

72 + 19 = ______ 453 + 48 = ______ 129 + 33 = ______

Subtract using shortcuts.

72 − 19 = ______ 453 − 48 = ______ 129 − 33 = ______

Use side 2 of your abacus to find the differences.

1. $\begin{array}{r} 8445 \\ -5372 \\ \hline \end{array}$

2. $\begin{array}{r} 3189 \\ -1734 \\ \hline \end{array}$

3. $\begin{array}{r} 7240 \\ -5808 \\ \hline \end{array}$

4. $\begin{array}{r} 2159 \\ -451 \\ \hline \end{array}$

5. $\begin{array}{r} 2573 \\ -1094 \\ \hline \end{array}$

6. $\begin{array}{r} 9062 \\ -5146 \\ \hline \end{array}$

---

7. $\begin{array}{r} 3079 \\ -1836 \\ \hline \end{array}$

8. $\begin{array}{r} 8625 \\ -6632 \\ \hline \end{array}$

9. $\begin{array}{r} 9162 \\ -4585 \\ \hline \end{array}$

10. $\begin{array}{r} 9468 \\ -8273 \\ \hline \end{array}$

11. $\begin{array}{r} 4883 \\ -1709 \\ \hline \end{array}$

12. $\begin{array}{r} 2728 \\ -1859 \\ \hline \end{array}$

Worksheet 9, Checking Subtraction by Adding Up

Name: ______________________

Date: ______________________

Warm-Up

Subtract.

$$\begin{array}{r} 143 \\ -29 \\ \hline \end{array} \qquad \begin{array}{r} 587 \\ -392 \\ \hline \end{array} \qquad \begin{array}{r} 847 \\ -598 \\ \hline \end{array}$$

Use adding up and check numbers to find the errors. Then correct the errors in the space below.

1. $\begin{array}{rr} 8449 & (7) \\ -\ 4177 & (1) \\ \hline 4372 & (7) \end{array}$

2. $\begin{array}{rr} 9427 & (4) \\ -\ 7921 & (1) \\ \hline 1596 & (3) \end{array}$

3. $\begin{array}{rr} 6966 & (0) \\ -\ 4138 & (7) \\ \hline 2838 & (2) \end{array}$

4. $\begin{array}{rr} 3443 & (5) \\ -\ 1088 & (8) \\ \hline 2355 & (7) \end{array}$

5. $\begin{array}{rr} 8194 & (4) \\ -\ 5532 & (6) \\ \hline 2552 & (2) \end{array}$

6. $\begin{array}{rr} 4912 & (7) \\ -\ 3134 & (2) \\ \hline 1778 & (5) \end{array}$

7. $\begin{array}{rr} 5129 & (8) \\ -\ 3186 & (0) \\ \hline 2043 & (0) \end{array}$

8. $\begin{array}{rr} 9095 & (5) \\ -\ 4697 & (8) \\ \hline 4498 & (7) \end{array}$

9. $\begin{array}{rr} 5660 & (8) \\ -\ 3496 & (4) \\ \hline 2164 & (4) \end{array}$

Subtract and check your work.

10. $\begin{array}{rr} 7797 & (\ \ ) \\ -\ 843 & (\ \ ) \\ \hline \end{array}$

11. $\begin{array}{rr} 6595 & (\ \ ) \\ -\ 5456 & (\ \ ) \\ \hline \end{array}$

12. $\begin{array}{rr} 6042 & (\ \ ) \\ -\ 3537 & (\ \ ) \\ \hline \end{array}$

13. $\begin{array}{rr} 2275 & (\ \ ) \\ -\ 757 & (\ \ ) \\ \hline \end{array}$

14. $\begin{array}{rr} 7890 & (\ \ ) \\ -\ 4362 & (\ \ ) \\ \hline \end{array}$

15. $\begin{array}{rr} 4175 & (\ \ ) \\ -\ 1802 & (\ \ ) \\ \hline \end{array}$

Name: ____________________

Date: ____________________

Warm-Up

Subtract these numbers. Use check numbers to check your work.

543 ( )	752 ( )	947 ( )
– 435 ( )	– 527 ( )	– 479 ( )

INFORMATION: A magic square is a special array of numbers. In a magic square, the sum of each row, column, and diagonal is the same. This sum is the *magic sum*.

Complete the magic squares below. Write the magic sums below them.

2	9	4
7	5	3
6	1	8

______

8		6
3	5	
4	9	2

______

16	3	2	13
5		11	
9	6		
4	15	14	1

______

9	6	3	16
4			5
14		8	11
7	12	13	

______

Complete the magic squares.

A

2	9	4
7		3
6	1	8

B

6	7	2
8	3	4

C

8	1	
3	5	
4	9	

D

	3	8
9		1
2	7	

E

6	1	
7	5	3
	9	

F

4		2
	5	
8		6

G

2		6
9		1
	3	

H

8	3	
1	5	

Name: ______________________________

Date: ______________________________

Warm-Up

Subtract these numbers. Use check numbers to check your work.

```
  872 ( )        734 ( )        429 ( )
– 728 ( )      – 347 ( )      – 294 ( )
```

1. Add 7 to each cell in the magic square to make a new array. Is the new array a magic square? ____

2	9	4
7	5	3
6	1	8

+ 7

9		

magic sum: ____ magic sum: ____

2. Subtract 3 to each cell in the magic square to make a new array. Is the new array a magic square? ______

6	1	8
7	5	3
2	9	4

– 3

3		

magic sum: ____ magic sum: ____

3. Multiply each cell in the magic square by 11 to make a new array. Is the new array a magic square? ____

4	9	2
3	5	7
8	1	6

× 11

44		

magic sum: _____ magic sum: _____

4. Add corresponding numbers in the two magic squares. Is the new array a magic square? ______

–1	3	–2
–1	0	1
2	–3	1

+

2	1	6
7	3	–1
0	5	4

=

1		

magic sum: ____ magic sum: ____ magic sum: ____

5. What is the relationship between the magic sum and the number in the middle cell?

____

____

6. Use the relationship you found at the left to complete these magic squares.

3		10
13	6	–1

13	3	5
		15

10	4	
	8	

Name: ______________________

Date: ______________________

Warm-Up

Complete these equations.

6 – 3 = _______ 5 – 6 = _______ –1 + 5 = _______

2 – 3 = _______ 9 – 11 = _______ –3 + 2 = _______

7 – 9 = _______ 1 – 5 = _______ –1 + 8 = _______

Complete the magic squares below.

A

7	12	1	14
	13	8	11
16	3		5
9		15	4

B

12	1	14	7
3	10	5	16
6	15	4	9

C

4	14	15	
9	7	6	
5	11	10	
16	2	3	

D

	3	4	13
10			0
17	4		–7
–7	13	0	14

E

	12	8	
15	6	10	3
14	7	11	2
4	9	5	

F

6			9
3	10	5	16
13	8	11	2
12		14	7

G

		–2	8
9	1	2	3
–2	7	5	5
–2	8		

H

1	14	14	
11	7		9
8		10	5
	2	3	15

1. Which two magic squares are reflections of each other? ____________

2. Which two magic squares are rotations of each other? ____________

3. Which three magic squares are not normal? ______________

4. What is special about the sum of the 4 corners? ______________________________

5. What is special about the sum of the inner squares? ___________________________

6. Top middle numbers + _________________________ = magic sum.

7. Left middle numbers + _________________________ = magic sum.

8. Add the corresponding cells in magic squares G and H. Write the sums in the square at the right. What is its magic sum? ____________

Name: ______________________

Date: ______________________

Warm-Up

Complete these equations.

27 – 21 = _______	114 – 115 = _______	10 – 30 = _______
27 – 29 = _______	128 – 130 = _______	40 – 50 = _______
27 – 30 = _______	497 – 500 = _______	200 – 300 = _______

Terry found another way to subtract. Terry's friends like it because they don't need to use any facts like 14 – 9. Here are some examples. What do you think?

87 – 49 = 40 – 2 = 38

516 – 394 = 200 – 80 + 2 = 122

7365 – 5468 = 2000 – 100 – 3 = 1897

Before you say, "I don't get it," let Terry explain it to you.

I start by subtracting the thousands.

Now hundreds—oops, not enough. So I write – 100.

Tens are easy: nothing to write.

7365 – 5468 = 2000 – 100 – 3 = 1897

Not enough ones. So I write – 3.

Then combine: 2000 – 100 = 1900 & 1900 – 3 = 1897.

Now you try it. Be sure to check your work.

91 – 57 = ______________________

371 – 191 = ______________________

809 – 437 = ______________________

4792 – 3285 = ______________________

3141 – 893 = ______________________

7040 – 2769 = ______________________

Name: ______________________________

Date: ______________________________

Warm-Up

Use Terry's way to subtract. Use check numbers if you like.

719 – 521 = ______________________________

462 – 293 = ______________________________

Terry's friends liked to subtract Terry's way, but they wanted to write the subtraction vertically. Here are some examples. What do you think?

54	852	6592
– 29	– 375	– 3718
30	500	3000
–5	–20	–200
25	–3	80
	477	–6
		2874

Here is Terry again, explaining it to you.

6592	6592	6592	6592	6592
– 3718	– 3718	– 3718	– 3718	– 3718
3000	3000	3000	3000	3000
	–200	–200	–200	–200
		80	80	80
			–6	–6
				2874

I start by subtracting the thousands.

Now hundreds—oops, not enough. So I write –200.

Tens are easy: 90 – 10 = 80.

Not enough ones. So I write –6.

Then combine: 3000 – 200 = 2800 and 80 – 6 = 74. The sum is 2874.

Now you try it. Be sure to check your work.

62	910	859	6076	3843	7083	6005
– 35	– 112	– 379	– 4059	– 3386	– 3259	– 1009

Name: ______________________

Date: ______________________

Write only the answers.

____

____

____

Write the answers.

582 + 69 = ______

87 + ______ = 200

(6 ÷ 3) + (6 ÷ 2) = ______

Add or subtract. Use check numbers.

9575 ( )	4763 ( )	9515 ( )	4012 ( )
+ 5592 ( )	+ 5251 ( )	– 5592 ( )	– 1802 ( )

Utah's population is two million nine hundred thousand eight hundred seventy-two. Underline the period names. Write the number using digits and commas. ______________________

Fill in the blanks.

3 × ____ = 24

8 × ____ = 64

7 × ____ = 14

____ × 11 = 44

____ × 9 = 54

6 × ____ = 24

2 × ____ = 14

Solve the problem.

Kendra wants to walk her dog for an hour. She has 25 minutes left to walk. How long has she walked so far?

Draw lines to match the expressions.

4 × 4	16 × 2
8 × 5	6 × 5 + 6 × 2
8 × 4	5 × 5
20 + 5	8 × 2
9 × 7	6 × 7 – 2
6 × 7	50 – 1
32 ÷ 4	9 × 6 + 2
7 × 7	70 – 7
8 × 7	2 × 2 × 2

Complete the magic square.

7	4	1	15
8		8	4
		8	
6	12		–1

Name: ______________________________

Date: ______________________________

Write only the answers.

____

____

____

Write the answers.

674 + 86 = ______

41 + ______ = 200

(8 ÷ 4) + (8 ÷ 2) = ______

Add or subtract. Use check numbers.

7864 ( )	9813 ( )	4603 ( )	7390 ( )
+ 3894 ( )	+ 9267 ( )	– 1597 ( )	– 3503 ( )

Hawaii's population is one million four hundred four thousand fifty-four. Underline the period names. Write the number using digits and commas. ______________________

Fill in the blanks.

2 × ____ = 16

6 × ____ = 36

7 × ____ = 56

11 × ____ = 66

____ × 4 = 16

8 × ____ = 56

____ × 9 = 36

Solve the problem.

Kevin is playing hockey for an hour. He has 35 minutes left to play. How long has he played already?

Draw lines to match the expressions.

9 × 9	8 × 3
7 × 5	10 × 8 – 8
6 × 4	90 – 9
8 × 8	6 × 6 – 1
9 × 8	80 – 16
50 + 6	7 × 8
40 ÷ 8	7 × 10 – 7
7 × 9	40 + 5
9 × 5	20 ÷ 4

Complete the magic square.

5	2	6	8
7		–1	
	3	16	4
11			–3

Name: ______________________

Date: ______________________

Solve these problems. Explain your answers.

1. The populations of some states are: California, 38,332,521; New York, 19,651,127; and Florida, 19,552,860. Does California have more people than New York and Florida together?

2. Peyton has $30.00 and is buying three food items costing $9.49, $7.97, and $11.53. Does Peyton have enough money?

3. When Skylar awoke in the morning, the temperature was –2°. By 3:00 p.m. the temperature is 10 degrees warmer. What is the temperature?

4. Avery is thinking of even numbers less than 50 that when divided by 7 have a remainder of 1. What are the numbers?

5. Tony and his family are traveling to visit relatives. The trip takes 2 hours and 40 minutes. If they leave at 9:10 and make two stops of 15 minutes each, what time will they arrive?

Name: ____________________

Date: ____________________

Warm-Up

Subtract. Use your favorite method. Check your work with check numbers or by adding up. numbers.

$$\begin{array}{r} 638\ (\ \ ) \\ -\ 429\ (\ \ ) \\ \hline \end{array} \qquad \begin{array}{r} 742\ (\ \ ) \\ -\ 289\ (\ \ ) \\ \hline \end{array}$$

Solve these number puzzles by putting in symbols such as: +, –, =, ×, ÷ to make equations.

1 4 2 2	2 3 1 6	6 2 1 2
____________	____________	____________
____________	____________	____________
		____________

Put <, > or = in the ellipses. Do not perform the calculations, but decide the answer by noticing patterns. Be prepared to justify your work.

a. 384 + 197 + 802 ◯ 381 + 196 + 799

b. 363 – 236 + 220 ◯ 363 – 236 + 234

c. 4765 – 2280 ◯ 4765 – 2995

d. 994 + 994 + 994 ◯ 994 × 4

e. 45 × 10 ◯ 45 × 5 × 2

f. 79 + 81 ◯ 80 × 2

g. 170 ÷ 2 ◯ 170 ÷ 5

h. 682 ÷ 3 ◯ 696 ÷ 3

i. 491 – 493 ◯ 0

j. 37 × 80 × 923 ◯ 923 × 80 × 37

Put <, > or = in the circles. Explain your reasons.

k. 14 × 68 ◯ 2 × 4 × 68 ____________________

l. 5716 – 378 ◯ 5720 – 382 ____________________

m. 6472 – 2959 ◯ 6472 – 2859 ____________________

n. 819 × 3 + 819 ◯ 819 × 4 ____________________

o. 712 ÷ 9 ◯ 712 ÷ 8 ____________________

Name: ______________________

Date: ______________________

Warm-Up

Find two solutions for these number puzzles. Make equations by putting in symbols such as: +, –, =, ×, ÷ .

1 2 3 4   5 1 1 6

______________   ______________

______________   ______________

Find the products. Write the partial products. Use your abacus for the first row.

$$\begin{array}{r} 365 \\ \times\ 9 \\ \hline \end{array} \qquad \begin{array}{r} 894 \\ \times\ 8 \\ \hline \end{array} \qquad \begin{array}{r} 1468 \\ \times\ 5 \\ \hline \end{array} \qquad \begin{array}{r} 1376 \\ \times\ 6 \\ \hline \end{array}$$

$$\begin{array}{r} 5608 \\ \times\ 8 \\ \hline \end{array} \qquad \begin{array}{r} 9876 \\ \times\ 9 \\ \hline \end{array} \qquad \begin{array}{r} 3579 \\ \times\ 4 \\ \hline \end{array} \qquad \begin{array}{r} \$51.86 \\ \times\ 7 \\ \hline \end{array}$$

Name: ____________________

Date: ____________________

Warm-Up

Find the products using partial products. Use your abacus if needed.

	3	8	2
		×	3

	6	3	9
		×	2

Use your abacus to do the following multiplications.

Multiply the number on the abacus by 2.

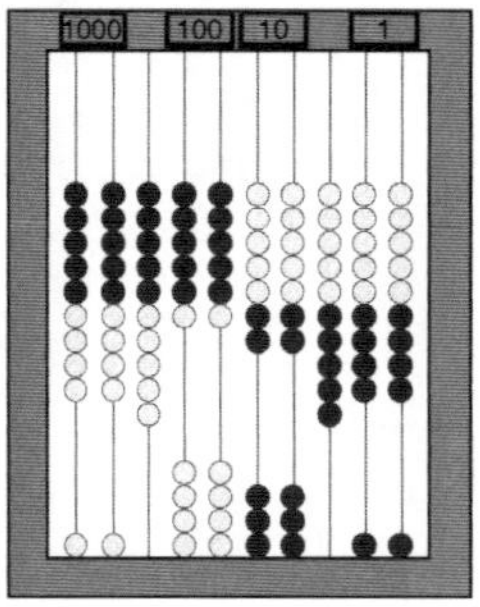

Multiply the number on the abacus by 6.

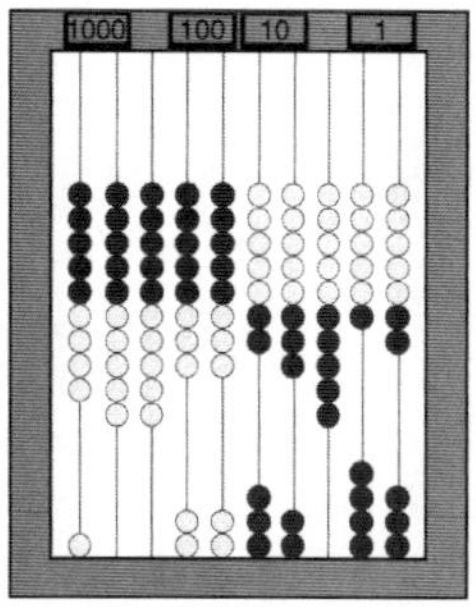

Multiply the number on the abacus by 4.

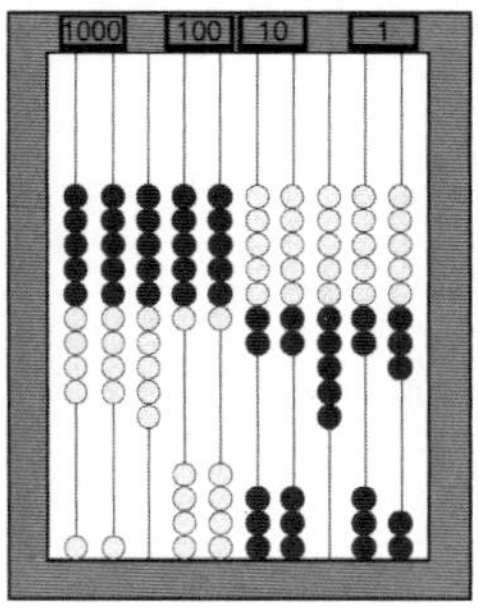

Multiply the number on the abacus by 8.

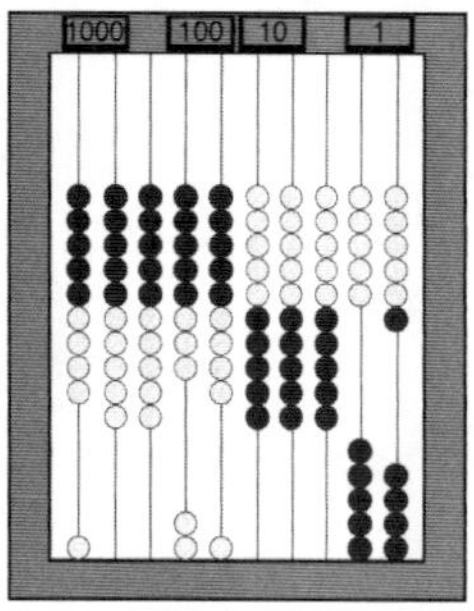

Multiply the number on the abacus by 7.

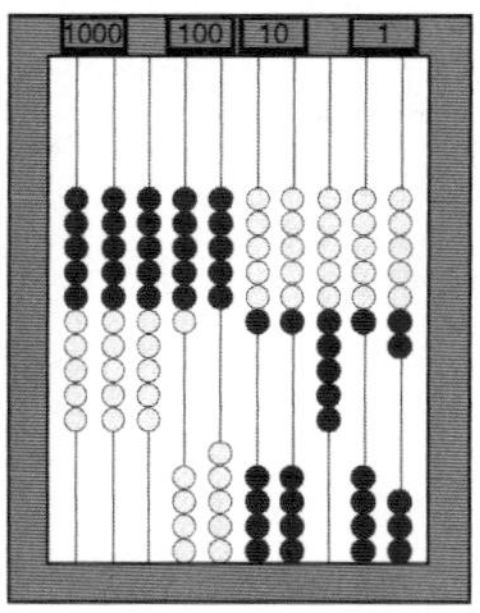

Multiply the number on the abacus by 5.

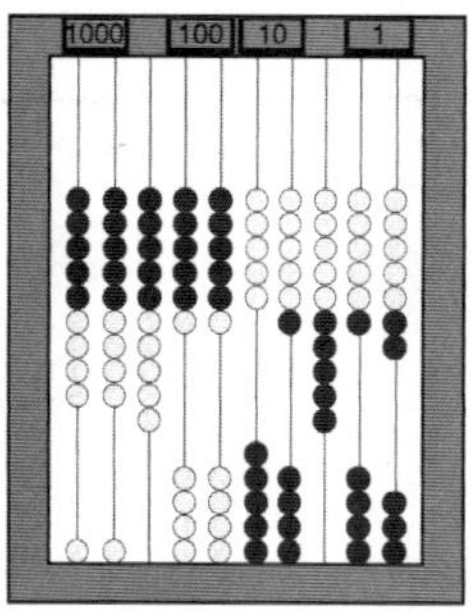

Name: ______________________

Date: ______________________

a.

Use partial products.

```
  3 2 1 3
  ×     3
```

Add.

```
  3 2 1 3
  3 2 1 3
+ 3 2 1 3
```

Multiply.

```
  3 2 1 3
  ×     3
```

b.

Use partial products.

```
  3 8 4 9
  ×     3
```

Add.

```
  3 8 4 9
  3 8 4 9
+ 3 8 4 9
```

Multiply.

```
  3 8 4 9
  ×     3
```

c.

Add. Use check numbers.

```
  4 5 1 9 ( )
  4 5 1 9 ( )
  4 5 1 9 ( )
  4 5 1 9 ( )
  4 5 1 9 ( )
+ 4 5 1 9 ( )
          ( )
```

Multiply. Use check numbers.

```
  4 5 1 9 ( )
  ×     6 ( )
          ( )
```

Multiply. Use check numbers.

```
  1 3 2 0 ( )
  ×     4 ( )
          ( )
```

```
  3 8 2 9 ( )
  ×     5 ( )
          ( )
```

```
  4 6 7 2 ( )
  ×     6 ( )
          ( )
```

```
  7 0 1 5 ( )
  ×     7 ( )
          ( )
```

```
  3 3 7 5 ( )
  ×     8 ( )
          ( )
```

```
  5 6 7 8 ( )
  ×     9 ( )
          ( )
```

When you find 6824 taken 7 times, which method is easiest—partial products, adding, or multiplying? Why? ______________________

____

____

Name: ______________________

Date: ______________________

Warm-Up

Complete the magic square.

6	–2		7
	8		
10			1
0	4	9	7

Find two solutions for this number puzzle. Make equations by putting in symbols such as: +, –, =, ×, ÷ .

1 2 2 4

____

____

Write equations and solve the following problems.

1. Sherry is 11 years old. Her grandfather is 7 times older than she is. How old is her grandfather? How much younger is Sherry than her grandfather?

2. The Confederation Bridge in Canada is 8 miles long. Lake Pontchartrain Causeway in Louisiana is 24 miles long. How many times longer is the causeway than the bridge? How much longer is the causeway than the bridge?

3. At birth an elephant weighs about 100 kg. When it is grown, it weighs 35 times as much. How much is that? How much weight did the baby gain growing up?

4. Harry is 4 ft tall. A foot is 12 times longer than an inch. How tall is Harry in inches?

Name: ______________________________

Date: ______________________________

Write only the answers.

____

____

____

Write the answers.

641 + 273 = ______

49 + ______ = 600

(12 ÷ 3) + (12 ÷ 2) = ______

Do the arithmetic. Use check numbers.

$$\begin{array}{r} 8374\ (\ \ ) \\ +\ 3558\ (\ \ ) \\ \hline \end{array} \qquad \begin{array}{r} 7000\ (\ \ ) \\ -\ 2001\ (\ \ ) \\ \hline \end{array} \qquad \begin{array}{r} 629\ (\ \ ) \\ \times 4\ (\ \ ) \\ \hline \end{array} \qquad \begin{array}{r} 4172\ (\ \ ) \\ \times 5\ (\ \ ) \\ \hline \end{array}$$

Multiply and add.

1 × 9 + 2 = ______ 12 × 9 + 3 = ______ 123 × 9 + 4 = ______ 1234 × 9 + 5 = ______

Solve the problems.

Sam earned \$7.75 on Tuesday and \$7.25 on Thursday. Then Sam bought some items. They cost 99¢, \$11.05, and \$2.15. How much money does Sam have left?

____

____

____

____

Steel is 8 times heavier than ice. If a cube of ice weighs 23 g, what would the cube weigh if it were made of steel?

____

____

____

The sun is one hundred forty-nine billion five hundred ninty-seven million eight hundred seventy thousand seven hundred meters from Earth. Write this number with numerals.

____

Name: ______________________________

Date: ______________________________

Round the populations of the three largest cities in England.

City in England	Population	Round to nearest thousand.
London	9,787,426	
Manchester	2,553,379	
Birmingham	2,440,986	

Fill in the table.

	Quotient	Remainder
41 ÷ 5		
41 ÷ 6		
41 ÷ 7		
41 ÷ 8		
41 ÷ 9		
41 ÷ 10		

Complete the magic squares.

6	4	
– 4	3	10
		0

1	3	2	6
5		3	–1
	–3	7	
6	7	0	

Solve the number puzzle two ways using the symbols, =, +, –, ×, ÷.

2 ___ 3 ___ 5 ___ 1

2 ___ 3 ___ 5 ___ 1

Name: ______________________

Date: ______________________

Write only the answers.

____

____

____

Write the answers.

783 + 156 = ______

51 + ______ = 700

(16 ÷ 8) + (16 ÷ 4) = ______

Do the arithmetic. Use check numbers.

7946 ( )	8000 ( )	847 ( )	5286 ( )
+ 4565 ( )	− 3002 ( )	× 4 ( )	× 5 ( )

Multiply and add.

1 × 8 + 1 = ______ 12 × 8 + 2 = ______ 123 × 8 + 3 = ______ 1234 × 8 + 4 = ______

Solve the problems.

Micky earned $5.35 on Tuesday, $6.50 on Wednesday, and $7.65 on Thursday. Then Micky bought 2 items. They cost $5.99 and $12.15. How much money does Micky have left?

Gold is about 19 times heavier than ice. If a block of ice weighs 8 g, what would the block weigh if it were gold?

The sun is two billion seven hundred ninty-five million eighty-four thousand eight hundred miles from Neptune. Write this number with numerals.

____

Name: ______________________

Date: ______________________

Round the populations of the three largest cities in Scotland.

City in Scotland	Population	Round to nearest thousand.
Glasgow	599,650	
Edinburgh	492,680	
Aberdeen	217,120	

Fill in the table.

	Quotient	Remainder
51 ÷ 5		
51 ÷ 6		
51 ÷ 7		
51 ÷ 8		
51 ÷ 9		
51 ÷ 10		

Complete the magic squares.

	0	8
10	3	– 4
	6	

1	4	6	– 1
4	0		3
	8		–2
0		2	10

Solve the number puzzle two ways using the symbols, =, +, –, ×, ÷.

6 ___ 1 ___ 2 ___ 3

6 ___ 1 ___ 2 ___ 3

Name: ______________________________ Date: ______________________________

Use your drawing tools to draw all lines.

Draw the rungs on the ladder with your T-square.

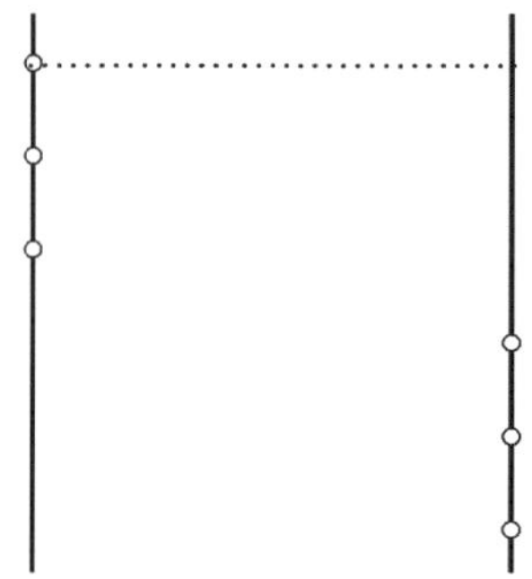

Draw horizontal lines to divide these figures in half using your T-square.

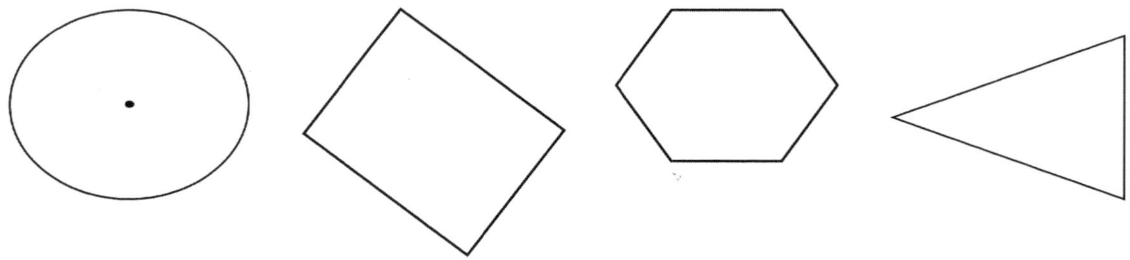

A diagonal is a line draw between two vertices (the points) in a polygon.
Draw the nine horizontal diagonals in the four polygons below.

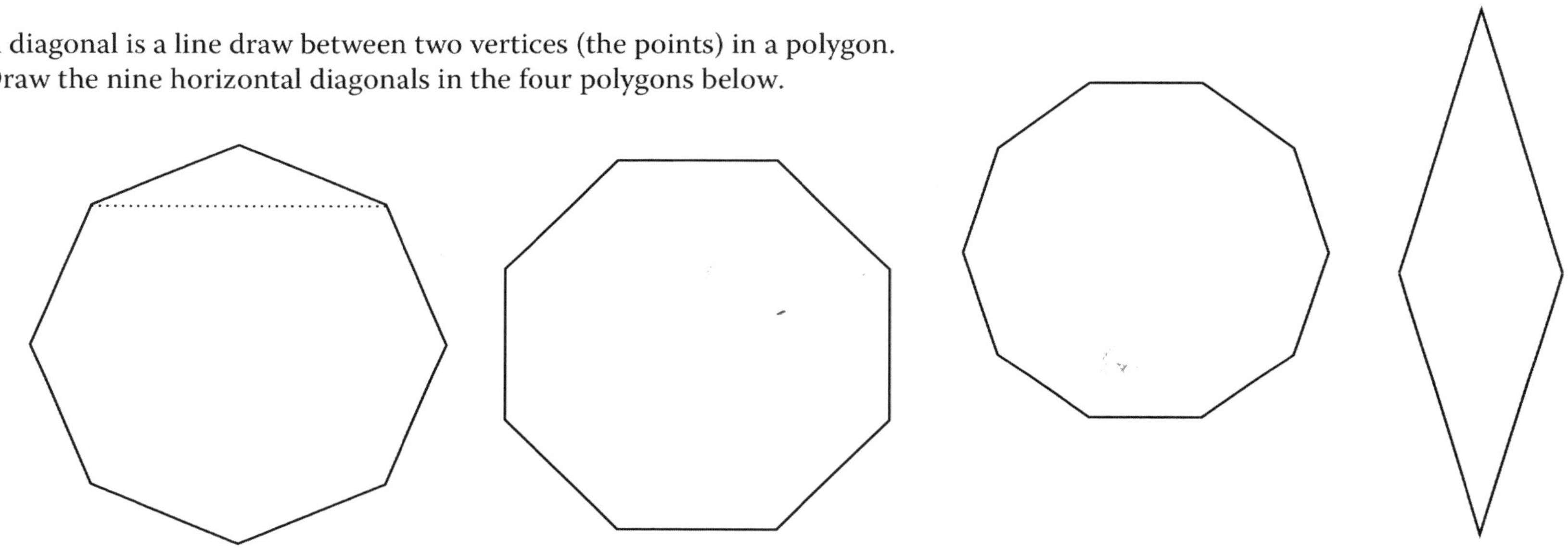

Name: ______________________ Date: ______________________

Use your drawing tools to draw all lines.

1. Draw the slats for the fence.

2. Complete the octagon.

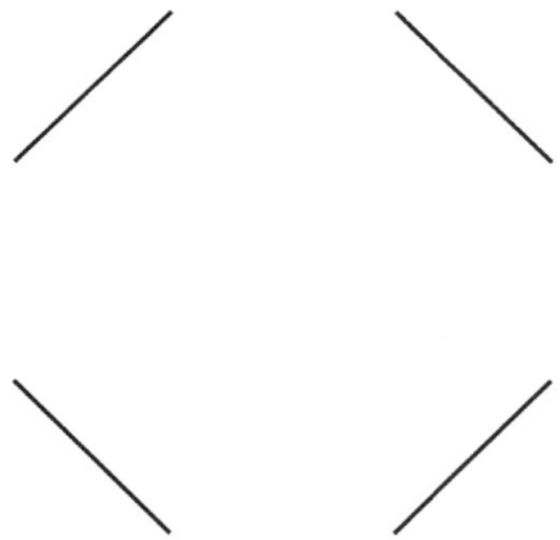

3. Draw both diagonals in the rhombuses.

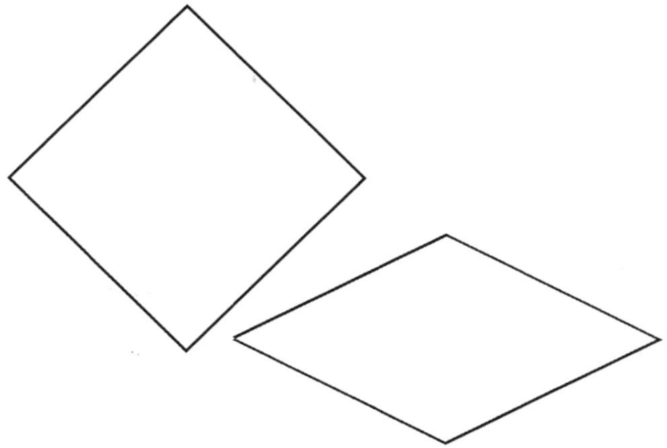

4. In the octagon draw the diagonals shown. Use your drawing tools with the 45 triangle.

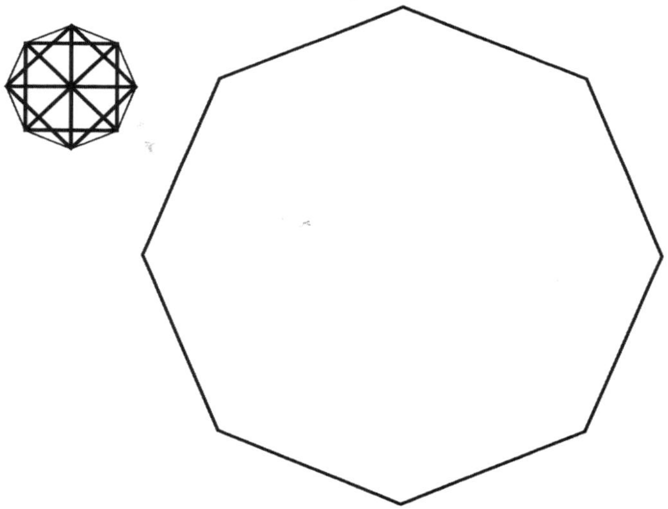

5. Draw all the diagonals in the hexagon using your drawing tools with the 30-60 triangle.

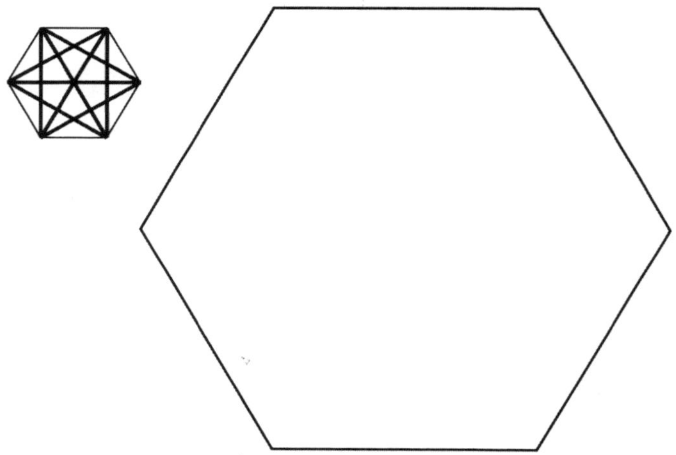

Name: ______________________________

Date: ______________________________

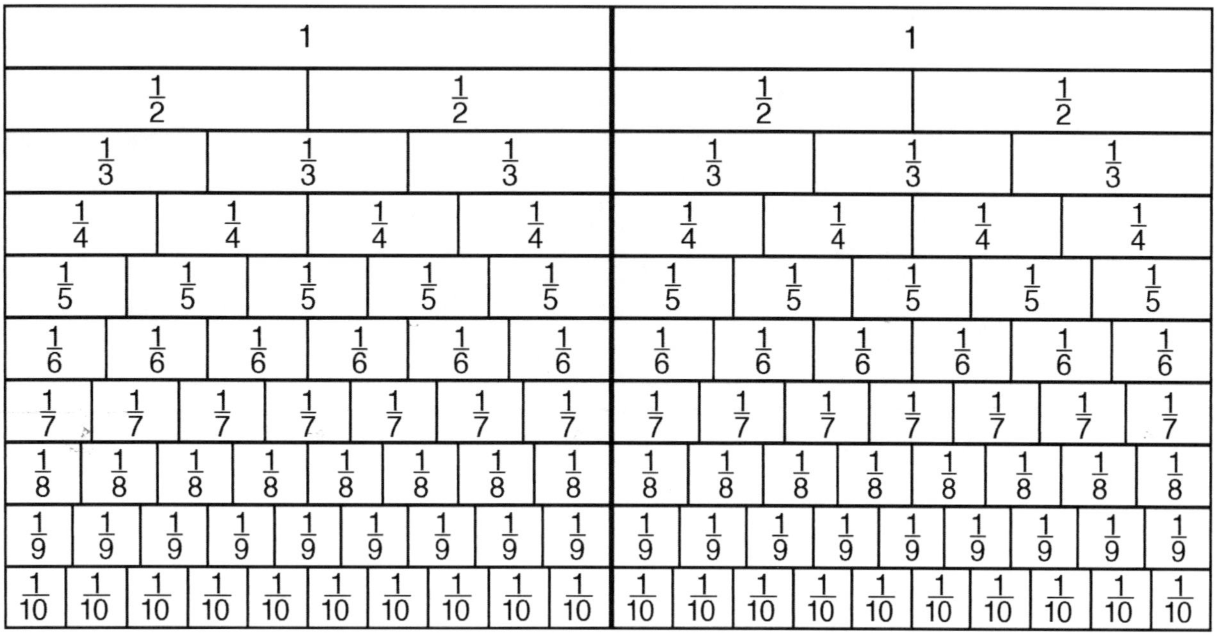

Use your drawing tools and the two fraction charts above to mark and label the following fractions: $\frac{1}{2}$, $\frac{1}{3}$, $\frac{2}{3}$, $1\frac{1}{4}$, $1\frac{2}{4}$, $1\frac{3}{4}$, and 2.

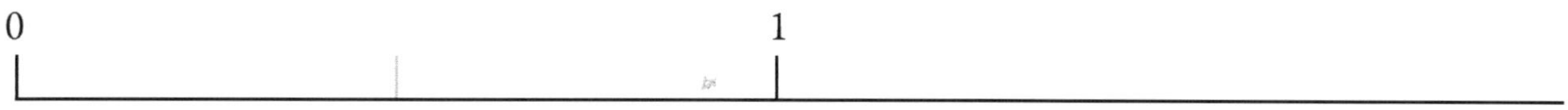

Use the fraction charts and drawing tools to find the fraction for each mark on the line. Write the fractions using lowest terms.

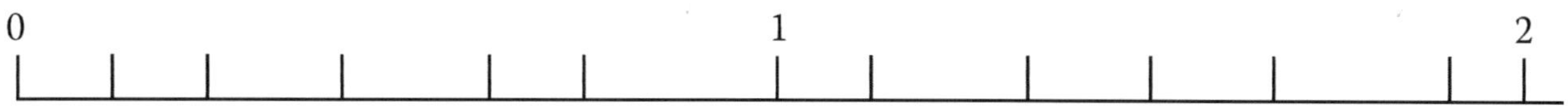

Write other fractions equivalent to the one at the left.

$\frac{4}{8}$ _____ _____ _____ _____

$\frac{3}{2}$ _____ _____ _____ _____

$\frac{2}{6}$ _____ _____

$\frac{2}{3}$ _____ _____

Name: ______________________________

Date: ______________________________

1. Use the T-square and 30-60 triangle to find and label the following fractions: $\frac{1}{2}$, $\frac{1}{4}$, $\frac{1}{8}$, and $\frac{1}{16}$.

2. Use the T-square and 45 triangle to divide the squares using diagonals into the fractions shown.

Halves Fourths Eighths Sixteenths

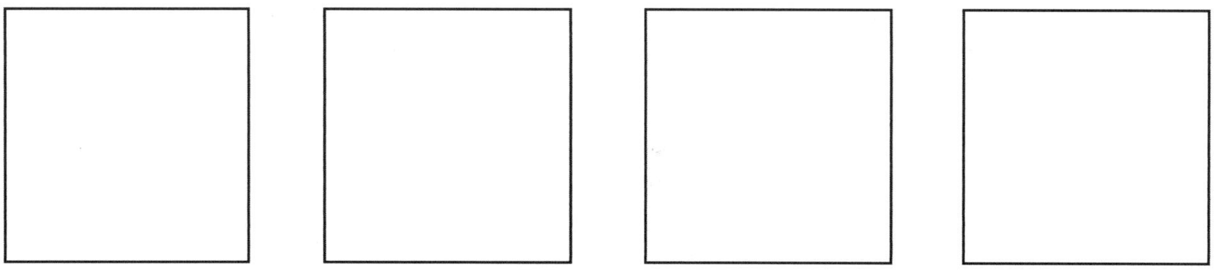

3. Use the T-square and 30-60 triangle to divide the triangles into the fractions shown.

Halves Fourths Eighths Sixteenths

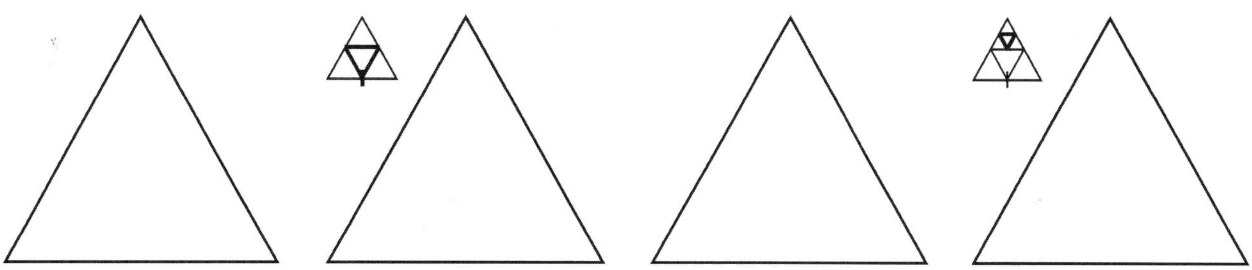

Name: ______________________

Date: ______________________

Warm-Up

Multiply.

2 7 ( )	4 6 ( )	9 4 ( )	1 2 9 ( )
× 3 ( )	× 8 ( )	× 6 ( )	× 5 ( )
( )	( )	( )	( )

For each problem, write the fraction of the rectangle that is crosshatched on the line.

1. Divide the rectangle into fourths and each fourth into halves. Crosshatch 3 pieces. ____

2. Divide the rectangle into halves and each half into thirds. Crosshatch 2 pieces. ____

3. Divide the rectangle into thirds and each third into thirds. Crosshatch 5 pieces. ____

4. Crosshatch one half of this rectangle. ____

   Also crosshatch one half of this rectangle and then divide each half into halves. ____

   Crosshatch one half of this rectangle and then divide each half into halves and each of those halves in half. ____

5. Crosshatch two thirds of this rectangle. ____

   Crosshatch two thirds of this rectangle. Divide each third in half. ____

   Crosshatch two thirds of this rectangle. Divide each third in half. Then divide each piece in half again. ____

6. Divide the rectangle into fourths and each fourth into thirds. Crosshatch one piece. ____

   Divide the rectangle into thirds and each third into fourths. Crosshatch one piece. ____

7. Divide the rectangle in half, but draw the line only halfway down. Divide the halves in half, but draw the lines only one fourth the way down. Divide the fourths in half, but draw the lines only one eighth the way down.

Name: ______________________

Date: ______________________

Warm-Up

Do the arithmetic. Use check numbers.

4982 ( )	4982 ( )	498 ( )	2894 ( )
+ 2894 ( )	− 2894 ( )	× 4 ( )	× 7 ( )

This rectangle is 1 inch long.

1. Draw a rectangle $1\frac{1}{2}$ inches long.
2. Draw a rectangle $1\frac{1}{4}$ inches long.
3. Draw a rectangle $1\frac{3}{4}$ inches long.
4. Draw a rectangle 2 inches long.
5. Draw a rectangle $\frac{1}{8}$ inch long.
6. Draw a rectangle $2\frac{1}{8}$ inches long.
7. Draw a rectangle $2\frac{1}{4}$ inches long.

From the cards shown on the right, choose the correct card to make each row below equal 1 and write the card in. Use your fraction chart if needed.

$\frac{1}{2}$ $\frac{1}{3}$ $\frac{1}{8}$ $\frac{1}{2}$ $\frac{7}{10}$ $\frac{3}{8}$ $\frac{2}{5}$ $\frac{1}{8}$ $\frac{1}{4}$

8. $\frac{7}{8}$ ☐

9. $\frac{1}{2}$ $\frac{1}{4}$ ☐

10. $\frac{2}{5}$ $\frac{1}{5}$ ☐

11. $\frac{1}{3}$ $\frac{1}{6}$ ☐

12. $\frac{1}{10}$ $\frac{1}{10}$ $\frac{3}{10}$ ☐

13. $\frac{1}{10}$ $\frac{1}{5}$ ☐

14. $\frac{1}{2}$ $\frac{1}{8}$ ☐

15. $\frac{1}{4}$ $\frac{1}{2}$ $\frac{1}{8}$ ☐

16. $\frac{1}{6}$ $\frac{1}{3}$ $\frac{1}{6}$ ☐

Worksheet 30–A, Review 3

Name: ______________________________

Date: ______________________________

Write only the answers.

____

____

____

Write the answers.

746 + 84 = ______

45 + ______ = 1000

$2\frac{2}{5} + 1\frac{3}{5} =$ ______

Do the arithmetic. Use check numbers.

24,479 ( )	9047 ( )	8529 ( )	4172 ( )
+ 37,548 ( )	– 3567 ( )	× 7 ( )	× 5 ( )

The population of Red Deer and the nearby area is 100,807. Write this number in words.

____

Write >, <, or = on the lines.

$\frac{1}{2}$ of 4 ____ $\frac{1}{2}$ of 2

$\frac{2}{4}$ ____ $\frac{1}{2}$

$1 + \frac{1}{2}$ ____ $1\frac{1}{2}$

$\frac{2}{3}$ ____ $\frac{2}{4}$

$\frac{3}{3}$ ____ $\frac{4}{4}$

$\frac{1}{6}$ ____ $\frac{5}{6}$

Solve the problem.

Aidan earned $16 on Tuesday and three times as much on Saturday. How much did Aidan earn on Saturday? How much did Aidan earn on both days?

Draw lines to match the expressions.

one half	4
3 ÷ 2	$\frac{1}{2}$
one half of 8	$\frac{2}{3}$
$\frac{1}{2}$ of $\frac{1}{2}$	$1\frac{1}{2}$
$\frac{1}{3} \times 2$	$\frac{1}{4}$
$\frac{1}{10} + \frac{9}{10}$	$\frac{7}{10}$
6 ÷ 2	1
$1 - \frac{3}{10}$	$\frac{6}{2}$

Crosshatch $\frac{2}{3}$ of the hexagon.

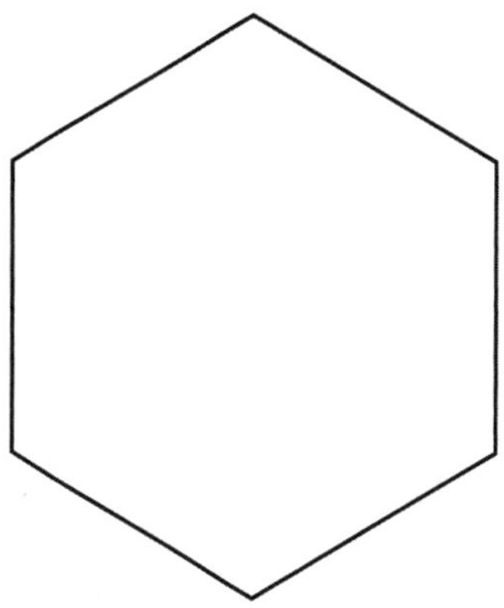

Name: ______________________________

Date: ______________________________

Write only the answers.

____

____

____

Write the answers.

$657 + 94 =$ ______

$63 +$ ______ $= 1000$

$1\frac{5}{6} + 2\frac{1}{6} =$ ______

Do the arithmetic. Use check numbers.

42,974 ( )	8703 ( )	2958 ( )	7251 ( )
+ 38,845 ( )	– 4596 ( )	× 8 ( )	× 6 ( )

The population of Markham, Ontario, is 301,709. Write this number in words.

____

Write >, <, or = on the lines.

$\frac{1}{2}$ of 3 ____ $\frac{1}{2}$ of 4

$\frac{1}{2}$ ____ $\frac{3}{6}$

$\frac{1}{2} + 1$ ____ $1\frac{1}{2}$

$\frac{3}{4}$ ____ $\frac{3}{6}$

$\frac{2}{2}$ ____ $\frac{8}{8}$

$\frac{1}{8}$ ____ $\frac{5}{8}$

Solve the problem.

James and his friends earned $24 on Friday and three times as much on Saturday. How much did they earn on Saturday? How much did they earn on both days?

Draw lines to match the expressions.

one third	3
$5 \div 2$	$\frac{1}{3}$
one half of 6	$\frac{2}{5}$
$\frac{1}{2}$ of $\frac{1}{3}$	$2\frac{1}{2}$
$\frac{1}{5} \times 2$	$\frac{1}{6}$
$\frac{3}{10} + \frac{7}{10}$	$\frac{9}{10}$
$7 \div 2$	1
$1 - \frac{1}{10}$	$\frac{7}{2}$

Crosshatch $\frac{3}{4}$ of the octagon.

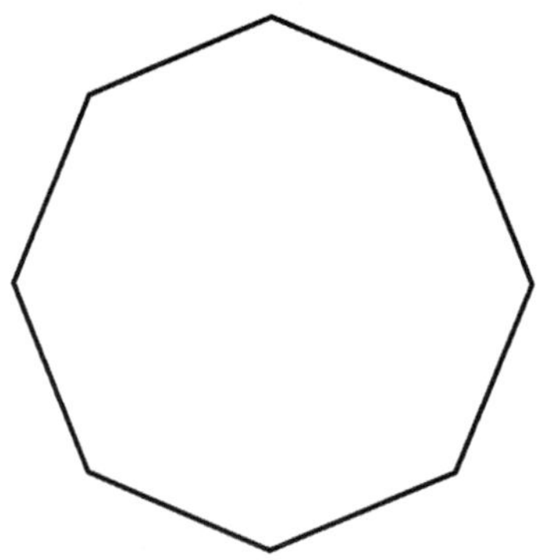

Name: ______________________________

Date: ______________________________

Warm-Up

Do the arithmetic. Use check numbers.

7268 ( )	7268 ( )	726 ( )	2687 ( )
+ 2687 ( )	− 2687 ( )	× 8 ( )	× 3 ( )

Solve these problems. Explain your answers.

1. Which rectangle has the greater perimeter?

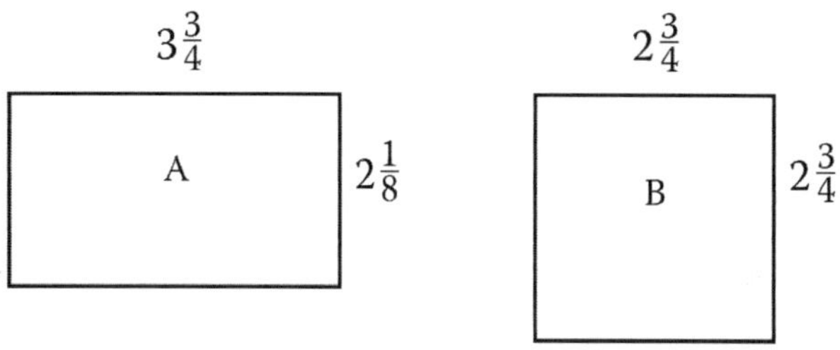

____

____

____

____

2. Sammy is making a craft project and has one foot of very expensive ribbon. Sammy cuts pieces $1\frac{1}{4}$, $1\frac{3}{4}$, $2\frac{5}{8}$, $2\frac{3}{4}$, $\frac{7}{8}$, and $1\frac{1}{2}$ inches long. How much ribbon is left? Explain your work.

____

____

____

____

Name: ____________________

Date: ____________________

Warm-Up

Do the arithmetic. Use check numbers.

5854 ( )	5854 ( )	585 ( )	4585 ( )
+ 4585 ( )	– 4585 ( )	× 6 ( )	× 7 ( )

Solve these problems.

1. The perimeter of each regular figure below is 13 cm. What is the length of the sides of each figure?

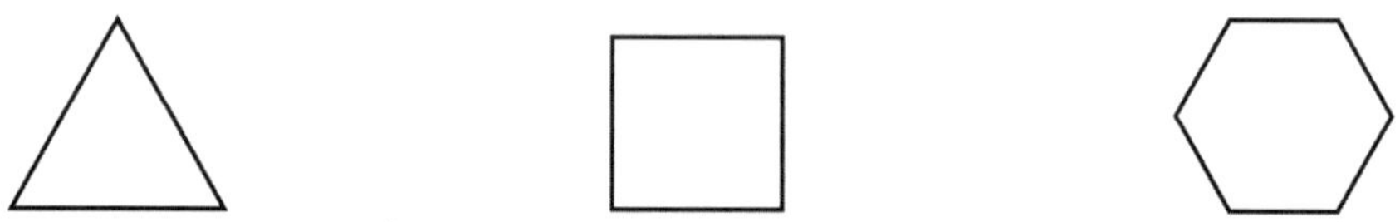

2. Find of the perimeter of the square and the regular octagon. Which has the greater perimeter?

Name: ______________________________

Date: ______________________________

Warm-Up

Do the arithmetic. Use check numbers.

$$\begin{array}{r} 7642\ (\ )\\ +\ 2467\ (\ )\\ \hline \end{array} \qquad \begin{array}{r} 7642\ (\ )\\ -\ 2467\ (\ )\\ \hline \end{array} \qquad \begin{array}{r} 764\ (\ )\\ \times\ 8\ (\ )\\ \hline \end{array} \qquad \begin{array}{r} 2467\ (\ )\\ \times\ 4\ (\ )\\ \hline \end{array}$$

1. Divide the rectangles by drawing vertical lines. Draw the lines freehand.

a. First find $\frac{1}{2}$ of the rectangle. Then crosshatch $\frac{1}{3}$ of $\frac{1}{2}$.

c. Find $\frac{1}{2}$ of the rectangle. Then crosshatch $\frac{3}{4}$ of $\frac{1}{2}$.

b. First find $\frac{1}{3}$ of the rectangle. Then crosshatch $\frac{1}{2}$ of $\frac{1}{3}$.

d. Find $\frac{3}{4}$ of the rectangle. Then crosshatch $\frac{1}{2}$ of $\frac{3}{4}$.

What pattern do you see? ______________________________

2. There are six rectangles below. Crosshatch $\frac{2}{3}$ of 6. How many thirds are crosshatched? ______

Now let's find $\frac{2}{3} \times 6$ a different way. Crosshatch $\frac{2}{3}$ of a rectangle. Repeat six times. Do not leave any gaps between the thirds. How many thirds are crosshatched? ______

Name: ____________________

Date: ____________________

Warm-Up

Do the arithmetic. Use check numbers.

6374 ( )	6374 ( )	637 ( )	4736 ( )
+ 4736 ( )	– 4736 ( )	× 4 ( )	× 8 ( )

1. What fraction are these inches divided into? _______

   Crosshatch the inch rectangles below.

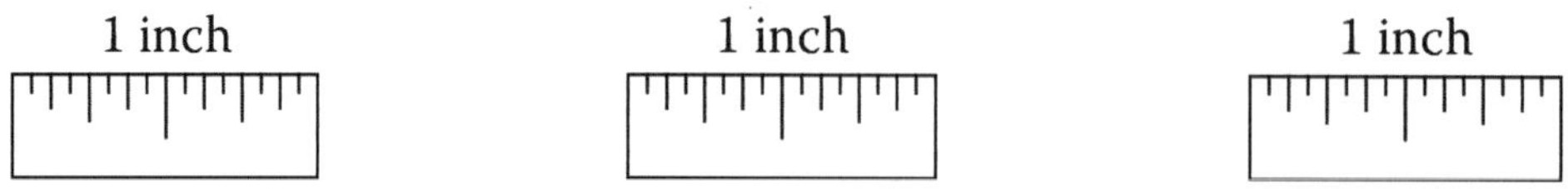

Crosshatch $\frac{3}{4}$. Crosshatch $\frac{3}{8}$. Crosshatch $\frac{5}{8}$.

2. Write the numbers below the ruler to show the inches.

   Draw lines inside the rectangle that are $1\frac{1}{4}$, $2\frac{3}{4}$, $4\frac{1}{2}$, and $5\frac{5}{8}$ inches long. Use the ruler below.

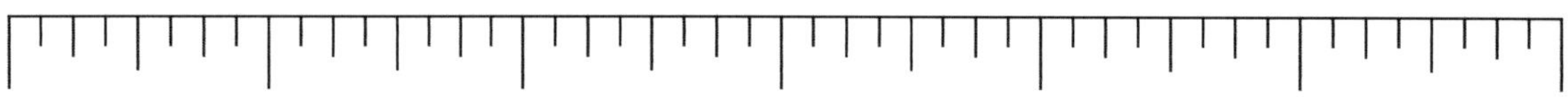

___ ___ ___ ___ ___ ___ ___

3. Use a ruler and find the perimeter of the two figures to the nearest eighth of an inch.

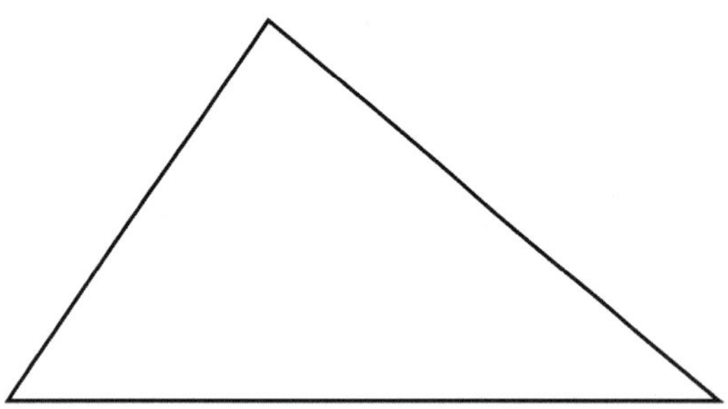

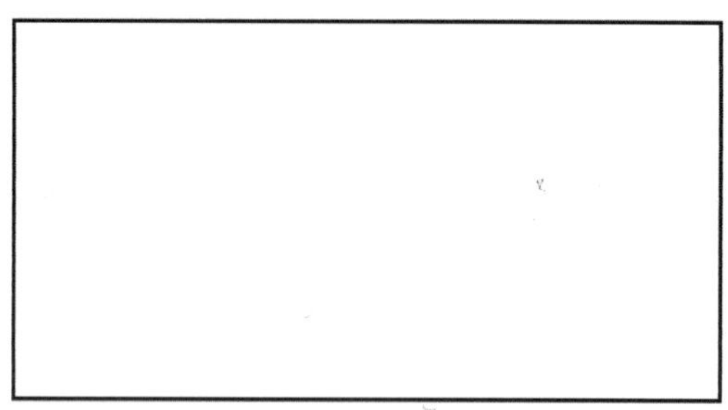

Name: ______________________________

Date: ______________________________

Write only the answers.

____

____

____

Write the answers.

$389 + 78 =$ _____

$61 +$ _____ $= 1000$

$2 - 1\frac{1}{3} =$ _____

Do the arithmetic. Use check numbers.

$$\begin{array}{r} 7432\ (\ \ ) \\ -\ 888\ (\ \ ) \\ \hline \end{array} \qquad \begin{array}{r} 7685\ (\ \ ) \\ \times\ 6\ (\ \ ) \\ \hline \end{array} \qquad \begin{array}{r} 6369\ (\ \ ) \\ \times\ 7\ (\ \ ) \\ \hline \end{array}$$

Add 387 + 10,793.

Write >, <, or = on the lines.

$\frac{1}{2}$ of 4 ____ $\frac{1}{2} \times 4$

$\frac{2}{5}$ ____ $\frac{4}{5}$

$1 + \frac{4}{7}$ ____ $1\frac{4}{7}$

$\frac{5}{6}$ ____ $\frac{4}{5}$

$\frac{7}{10}$ ____ $\frac{7}{8}$

$\frac{11}{8}$ ____ $1\frac{3}{8}$

Solve the problem.

Willie walked around a rectangular block. The north side of the block was 486 ft. The west side was twice this distance. How far did Willie walk?

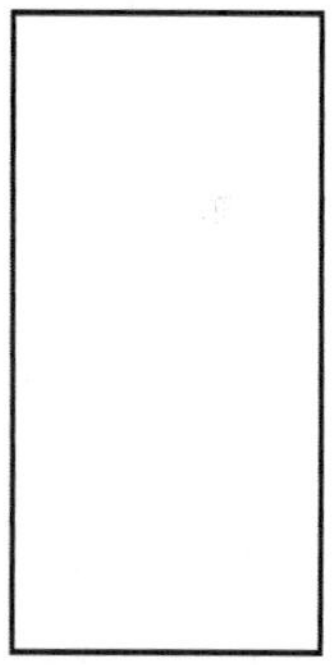

Circle the improper fractions.

$\frac{9}{8}$ $\frac{7}{1}$ $\frac{2}{3}$ $\frac{6}{6}$ $\frac{12}{13}$ $\frac{5}{6}$

Put these fractions in order from least to greatest.

$\frac{5}{4}$ $\frac{7}{8}$ $\frac{3}{5}$ $\frac{1}{3}$ $\frac{17}{8}$ $\frac{1}{4}$

____

On the fraction charts below, crosshatch $1\frac{1}{2}$, $\frac{2}{3}$, $\frac{7}{4}$, and $\frac{5}{5}$.

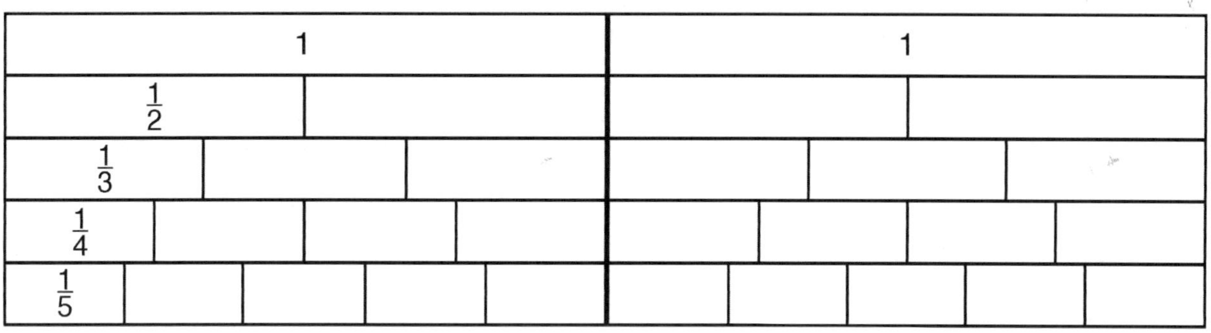

Name: ______________________

Date: ______________________

Write only the answers.

____

____

____

Write the answers.

$576 + 85 =$ _____

$73 +$ _____ $= 1000$

$2 - 1\frac{1}{4} =$ _____

Do the arithmetic. Use check numbers.

$$\begin{array}{r} 6345 \;(\;\;) \\ -\;548 \;(\;\;) \\ \hline \end{array} \qquad \begin{array}{r} 5867 \;(\;\;) \\ \times\;7 \;(\;\;) \\ \hline \end{array} \qquad \begin{array}{r} 7849 \;(\;\;) \\ \times\;6 \;(\;\;) \\ \hline \end{array}$$

Add 268 + 23,564.

Write >, <, or = on the lines.

$\frac{1}{3} \times 6$ ____ $\frac{1}{3}$ of 6

$\frac{5}{8}$ ____ $\frac{3}{8}$

$2 + \frac{5}{7}$ ____ $2\frac{5}{7}$

$\frac{7}{8}$ ____ $\frac{8}{9}$

$\frac{3}{10}$ ____ $\frac{3}{8}$

$\frac{13}{8}$ ____ $1\frac{5}{8}$

Solve the problem.

Matt's dog, Barky, ran around her pen. The east side of the pen was 57 ft. The south side was three times this distance. How far did Barky run?

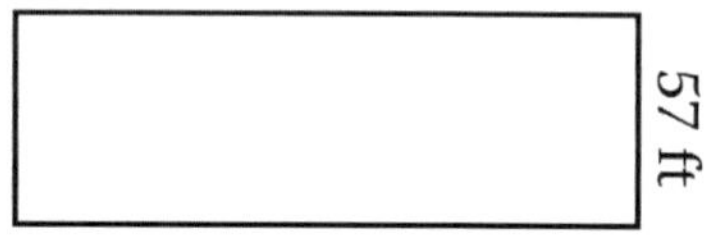

Circle the improper fractions.

$\frac{7}{8} \quad \frac{7}{3} \quad \frac{2}{1} \quad \frac{6}{6} \quad \frac{1}{5} \quad \frac{7}{6}$

Put these fractions in order from least to greatest.

$\frac{4}{4} \quad \frac{1}{8} \quad \frac{4}{7} \quad \frac{2}{5} \quad \frac{15}{8} \quad \frac{1}{6}$

____

On the fraction charts below, crosshatch $\frac{2}{2}$, $\frac{4}{3}$, $\frac{3}{4}$, and $\frac{7}{5}$.

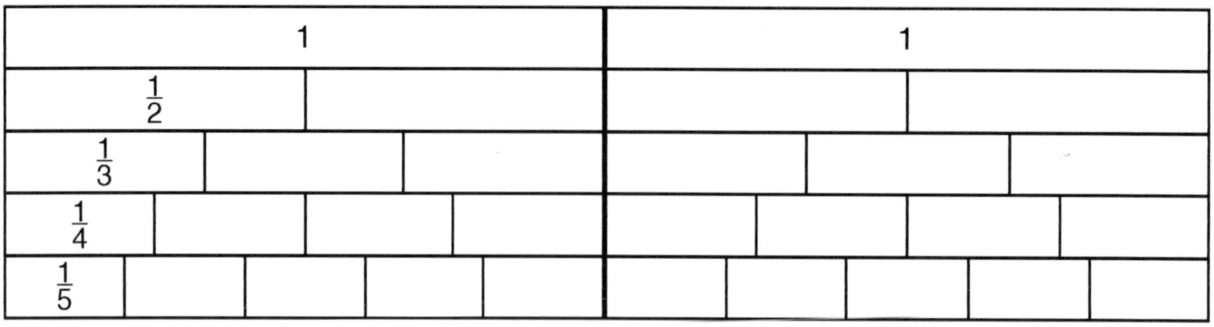

Name: ____________________

Date: ____________________

Warm-Up

Do the arithmetic. Use check numbers.

6532 ( )	6532 ( )	653 ( )	2356 ( )
+ 2356 ( )	– 2356 ( )	× 3 ( )	× 4 ( )

Multiply horizontally.

412 × 10 = ______ 720 × 10 = ______ 8397 × 10 = ______

86 × 2 = ______ 346 × 2 = ______ 5976 × 2 = ______

86 × 20 = ______ 346 × 20 = ______ 5976 × 20 = ______

217 × 4 = ______ 706 × 5 = ______ 4231 × 3 = ______

217 × 40 = ______ 706 × 50 = ______ 4231 × 30 = ______

Multiply vertically.

212	212	143	143	3502	3502
×2	×20	×3	×30	×4	×40

365	365	4321	4321	2432	2432
×5	×50	×6	×60	×7	×70

Name: ______________________________

Date: ______________________________

1. Find the products by using your previous answers wherever possible.

$$\begin{array}{r} 63 \\ \times\ 5 \\ \hline \end{array} \qquad \begin{array}{r} 63 \\ \times\ 30 \\ \hline \end{array} \qquad \begin{array}{r} 63 \\ \times\ 35 \\ \hline \end{array}$$

$$\begin{array}{r} 825 \\ \times\ 6 \\ \hline \end{array} \qquad \begin{array}{r} 825 \\ \times\ 50 \\ \hline \end{array} \qquad \begin{array}{r} 825 \\ \times\ 56 \\ \hline \end{array}$$

$$\begin{array}{r} 3674 \\ \times\ 1 \\ \hline \end{array} \qquad \begin{array}{r} 3674 \\ \times\ 80 \\ \hline \end{array} \qquad \begin{array}{r} 3674 \\ \times\ 81 \\ \hline \end{array}$$

$$\begin{array}{r} 9062 \\ \times\ 7 \\ \hline \end{array} \qquad \begin{array}{r} 9062 \\ \times\ 20 \\ \hline \end{array} \qquad \begin{array}{r} 9062 \\ \times\ 27 \\ \hline \end{array}$$

2. Multiply.

$$\begin{array}{r} 81 \\ \times\ 52 \\ \hline \end{array} \qquad \begin{array}{r} 143 \\ \times\ 33 \\ \hline \end{array} \qquad \begin{array}{r} 572 \\ \times\ 64 \\ \hline \end{array} \qquad \begin{array}{r} 2927 \\ \times\ 81 \\ \hline \end{array}$$

3. How many days are in the following number of years? Ignore leap years.

2 yr   9 yr   10 yr   55 yr   26 yr

3 6 5

Name: ______________________________

Date: ______________________________

Warm-Up

Multiply. Use check numbers to check your work.

$$\begin{array}{r} 43\,(\ ) \\ \times\ 27\,(\ ) \\ \hline (\ ) \end{array} \qquad \begin{array}{r} 287\,(\ ) \\ \times\ 45\,(\ ) \\ \hline (\ ) \end{array} \qquad \begin{array}{r} 8263\,(\ ) \\ \times\ 58\,(\ ) \\ \hline (\ ) \end{array}$$

Find the factor pairs of the numbers below. Then list the factors.

Factor pairs for 12: 1 & 12 __________ __________

Factors of 12: 1 2 ____ ____ ____ ____

Factor pairs for 10: __________ __________

Factors of 10: ____ ____ ____ ____

Factor pairs for 11: __________

Factors of 11: ____ ____

Factor pairs for 18: __________ __________ __________

Factors of 18: ____ ____ ____ ____ ____ ____

Factor pairs for 25: __________ __________

Factors of 25: ____ ____ ____

Factor pairs for 56: __________ __________ __________ __________

Factors of 56: ____ ____ ____ ____ ____ ____ ____ ____

Factor pairs for 48: __________ __________ __________ __________ __________

Factors of 48: ____ ____ ____ ____ ____ ____ ____ ____ ____ ____

Factor pairs for 100: __________ __________ __________ __________ __________

Factors of 100: ____ ____ ____ ____ ____ ____ ____ ____ ____

Name: ____________________

Date: ____________________

Warm-Up

Multiply.

```
  8 2 3 ( )          4 5 3 9 ( )
×   6 7 ( )            × 8 3 ( )
        ( )                  ( )
```

What are the factors of 7? ____ ____ factors of 6? ____ ____ ____ ____ factors of 2? ____ ____

A number having only two factors is called a *prime number.*

Is 7 a prime number? ______ Is 6 a prime number? ______ Is 2 a prime number? ______

The number 1 has only one factor, so 1 is not a prime number.

A number having more than two factors is called a *composite number.* "Composite" means composed of parts.

Is 7 a composite number? ______ Is 6 a composite number? ______

Write "true" or "false" for each statement.

_________ A prime number has only two factors.

_________ A composite number always has an even number of factors.

_________ Two is the only even prime number.

_________ The number one is neither a prime number nor a composite number.

How do you know that 8, 10, 16, and 98 are not primes? ____________________

How do you know that 15, 30, and 85 are not primes? ____________________

How do you know that 12, 51, and 84 are not primes? ____________________

How do you know that 14, 28, and 63 are not primes? ____________________

Use your calculator to help you find the answers.

Is 97 a prime number? ______

Is 307 a prime number? ______

Is 667 a prime number? ______

Name: ______________________________

Date: ______________________________

Warm-Up

Multiply.

$$\begin{array}{r} 382\,(\;) \\ \times\;49\,(\;) \\ \hline \\ \\ (\;) \end{array} \qquad \begin{array}{r} 1786\,(\;) \\ \times\;25\,(\;) \\ \hline \\ \\ (\;) \end{array}$$

Use the Sieve of Eratosthenes (air-uh-TOSS-the-knees) algorithm to find the prime numbers up to 100. Cross out the multiples of primes, but not the primes themselves. Circle the prime numbers.

1	2	3	4	5	6	7	8	9	10
11	12	13	14	15	16	17	18	19	20
21	22	23	24	25	26	27	28	29	30
31	32	33	34	35	36	37	38	39	40
41	42	43	44	45	46	47	48	49	50
51	52	53	54	55	56	57	58	59	60
61	62	63	64	65	66	67	68	69	70
71	72	73	74	75	76	77	78	79	80
81	82	83	84	85	86	87	88	89	90
91	92	93	94	95	96	97	98	99	100

List the prime numbers:

____ ____ ____ ____ ____
____ ____ ____ ____ ____
____ ____ ____ ____ ____
____ ____ ____ ____ ____
____ ____ ____ ____ ____

Goldbach Conjecture: Even numbers > 2 = sum of two primes.

Complete the equations with prime numbers.

12 = _____ + _____

18 = _____ + _____ 18 = _____ + _____

32 = _____ + _____ 32 = _____ + _____

40 = _____ + _____ 40 = _____ + _____ 40 = _____ + _____

58 = _____ + _____ 58 = _____ + _____ 58 = _____ + _____ 58 = _____ + _____

Name: ____________________

Date: ____________________

Warm-Up

In February, 2013, the largest known prime number had 17,425,170 digits. Then in January, 2016, a larger prime was found with 4,913,448 more digits. How many digits does the newest prime have? ____________

With five thousand digits fitting on an ordinary sheet of paper, the prime would take about 4400 sheets of paper. How many books does this make if each book has 100 pages? ________

Solve these problems.

1. Valerie has $23. She is buying books, which cost $5 each. How many books can she buy? What is her change?

2. Van is dividing 48 sheets of construction paper equally among 4 boys and 5 girls. How many will each child get? How many sheets of paper will be left?

3. Victor wants to buy 30 small items. They are sold 8 to a package. Each package costs $9. What will the items cost? How many extras will there be?

4. Violet is making a square pen for a pet rabbit. She has 14 m of fencing. What is the largest pen she can make? How much fencing will remain?

Worksheet 42, Dividing 4-Digit Numbers

Name: ______________________________

Date: ______________________________

Divide using side 2 of your abacus.

2)8476	2)6240	2)2850	2)6728
2)3464	2)9146	2)7362	2)4554

Divide using side 2 of your abacus.

3)6742	4)9384	7)8386	8)9759
9)9936	5)7304	6)6546	3)4622
5)4280	4)1824	3)1812	7)3760

Divide. Use the abacus only if you get stuck.

2)6846	2)9874	5)7585	7)8680
3)4023	3)4024	5)4280	5)4283
9)2886	6)5948	5)9562	8)7095

Name: ______________________

Date: ______________________

Warm-Up

Divide. Use check numbers to check your answers.

$2\overline{)6453}$ $3\overline{)6453}$ $5\overline{)6453}$

Multiply the top number by 2, then 3, up to the number shown. Next divide by 2, then 3, and so on, up to the highest number you multiplied by. Use check numbers.

2	3	4
× 2 ( )	× 2 ( )	× 2 ( )
× 3 ( )	× 3 ( )	× 3 ( )
× 4 ( )	× 4 ( )	× 4 ( )
× 5 ( )	× 5 ( )	× 5 ( )
× 6 ( )	× 6 ( )	× 6 ( )
( )	× 7 ( )	× 7 ( )
( )	( )	× 8 ( )
( )	( )	( )
( )	( )	( )
	( )	( )
	( )	( )
		( )
		( )
		( )

Write out in words the greatest number on this page.

____

____

Name: ______________________________

Date: ______________________________

Write only the answers.

____

____

____

Write the answers.

$641 + 273 =$ ______

$2\frac{2}{3} + 1\frac{1}{3} =$ ______

$(42 \div 6) + (42 \div 42) =$ ______

Do the arithmetic. Use check numbers.

Add: 467 + 14,829.

Subtract: 9037 – 288.

$$\begin{array}{r} 2676\ (\ \ ) \\ \underline{\times 25\ (\ \ )} \end{array}$$

An astronomical unit is the distance of the Earth to the sun, which is 149,597,870,700 meters. Round this number to billions and to millions.

______________________________ ______________________________

Solve the problems.

Which rectangle has the greater perimeter?

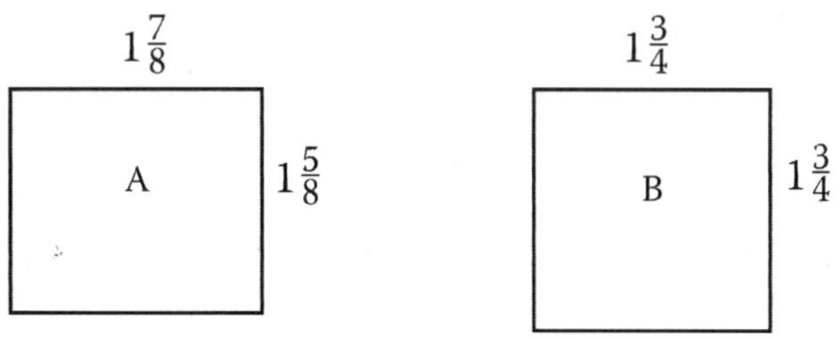

Use a ruler to measure the triangle below in centimeters. Then find the perimeter.

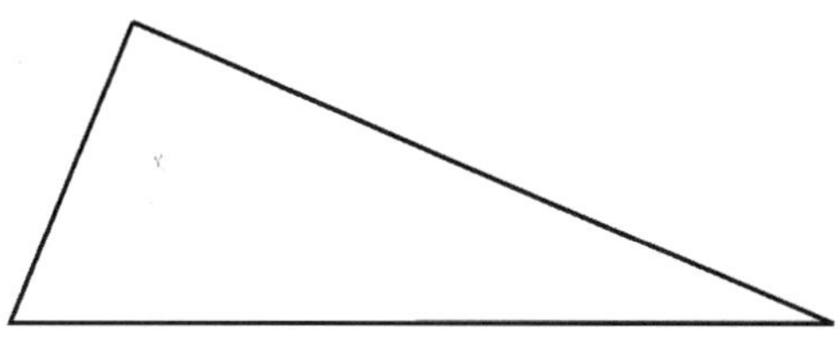

A baby giraffe at birth is 6 ft tall. A fully grown male giraffe is three times taller. How tall is a male giraffe? What is the difference in height between the baby and the adult?

Name: ______________________________

Date: ______________________________

Find the factor pairs of the numbers below. Then list the factors.

Factor pairs for 20: ________ ________ ________

Factors of 20: ____ ____ ____ ____ ____ ____

Factor pairs for 16: ________ ________ ________

Factors of 16: ____ ____ ____ ____ ____

Name the prime numbers between 6 and 12. __________

Multiply 5 by 5, 6, and 7. Then divide by 5, 6, and 7.

$$
\begin{array}{rl}
5 & \\
\times\ 5 & \\
\hline
 & (\ \ ) \\
\times\ 6 & \\
\hline
 & (\ \ ) \\
\times\ 7 & \\
\hline
 & (\ \ ) \\
 & (\ \ ) \\
 & (\ \ )
\end{array}
$$

Put these fractions in order from least to greatest.

$\frac{1}{2}$ $\frac{7}{8}$ $\frac{9}{10}$ $\frac{2}{3}$ $\frac{4}{3}$ $\frac{1}{4}$

____

Write >, <, or = on the lines.

$\frac{1}{2}$ of 6 ____ $\frac{1}{2} \times 6$

$\frac{9}{10}$ ____ $\frac{7}{10}$

$1 + \frac{4}{5}$ ____ $1\frac{4}{5}$

$\frac{5}{6}$ ____ $\frac{7}{8}$

$\frac{1}{2}$ ____ $\frac{3}{5}$

$\frac{11}{6}$ ____ $1\frac{1}{6}$

$\frac{1}{4}$ of 12 ____ 4

What fraction is shaded?

____
____
____
____
____

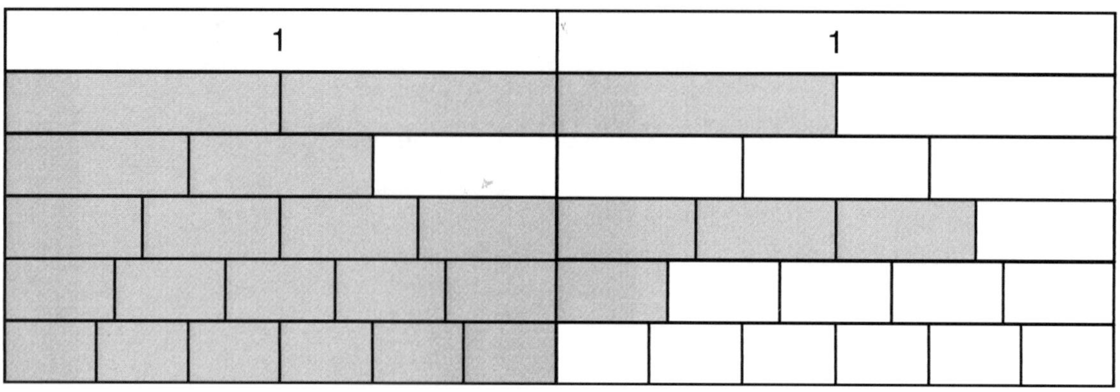

Name: ____________________

Date: ____________________

Write only the answers.

____

____

____

Write the answers.

$726 + 184 =$ ______

$4\frac{1}{5} + 1\frac{4}{5} =$ ______

$(42 \div 7) + (42 \div 42) =$ ______

Do the arithmetic. Use check numbers.

Add: 582 + 23,609.

Subtract: 8403 – 788.

$$\begin{array}{r} 4628\ (\ \ ) \\ \underline{\times 35\ (\ \ )} \end{array}$$

The distance around the Earth at the equator is 1,577,727,360 inches.

Round this number to millions and to thousands.

____________________ ____________________

Solve the problems.

Which rectangle has the greater perimeter?

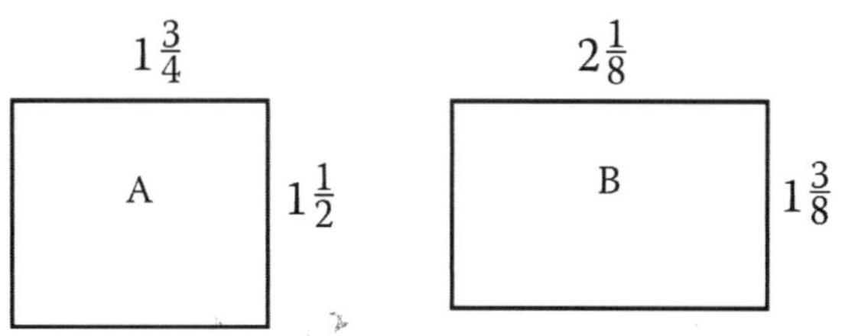

Use a ruler to measure the triangle below in centimeters. Then find the perimeter.

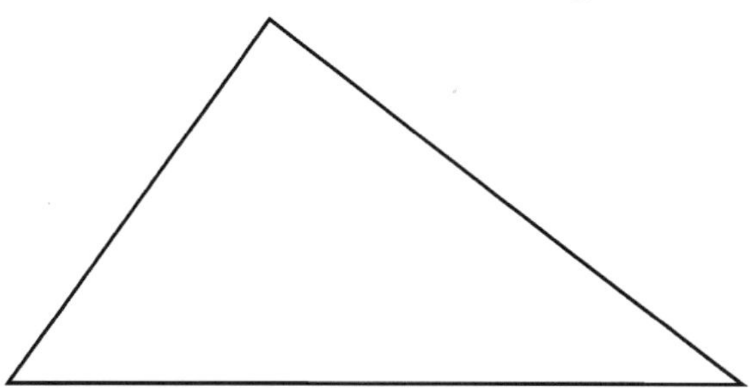

Hudson weighed 7 lb when he was born. Now he weighs 20 times that weight. What does he weight now? How much more does he weigh now than when he was born?

Name: ____________________

Date: ____________________

Find the factor pairs of the numbers below. Then list the factors.

Factor pairs for 24: ______ ______ ______ ______

Factors of 24: ____ ____ ____ ____ ____ ____ ____ ____

Factor pairs for 18: ______ ______ ______

Factors of 18: ____ ____ ____ ____ ____ ____

Name the prime numbers between 4 and 10. ______

Multiply 7 by 5, 6, and 7. Then divide by 5, 6, and 7.

$$\begin{array}{r} 7 \\ \times\ 5 \\ \hline \\ (\quad) \\ \times\ 6 \\ \hline \\ (\quad) \\ \times\ 7 \\ \hline \\ (\quad) \\ (\quad) \\ (\quad) \end{array}$$

Put these fractions in order from least to greatest.

$\frac{2}{5}$ $\frac{1}{8}$ $\frac{9}{10}$ $\frac{1}{2}$ $\frac{3}{5}$ $\frac{5}{4}$

____

Write >, <, or = on the lines.

$\frac{1}{4}$ of 4 ____ $\frac{1}{4} \times 4$

$\frac{9}{7}$ ____ $\frac{7}{7}$

$1\frac{5}{8}$ ____ $1 + \frac{5}{8}$

$\frac{5}{8}$ ____ $\frac{5}{7}$

$\frac{1}{2}$ ____ $\frac{4}{7}$

$\frac{11}{6}$ ____ $1\frac{5}{6}$

$\frac{1}{6}$ of 12 ____ 2

What fraction is shaded?

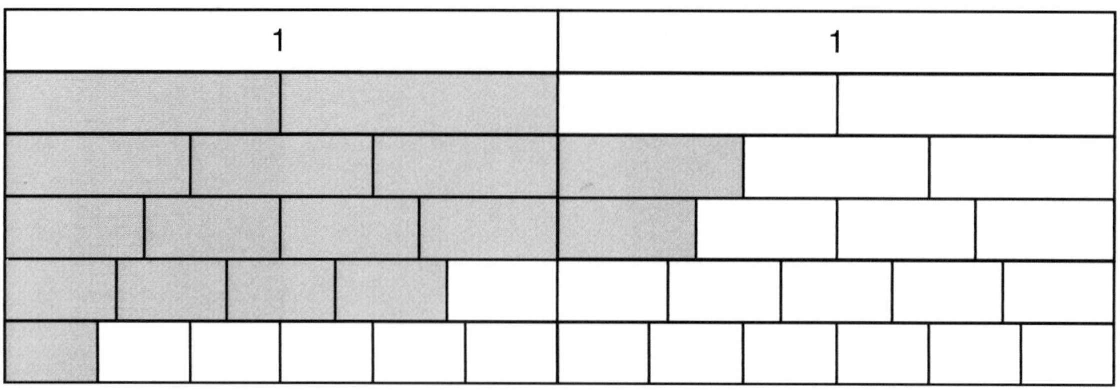

____
____
____
____
____

Worksheet 46, Working with Tenths

Name: ____________________

Date: ____________________

Measure and find the perimeter of the following figures.
Use tenths of an inch.

1. 

2. 

Measure and find the perimeter of the following regular figures.
Use tenths of a centimeter.

3. 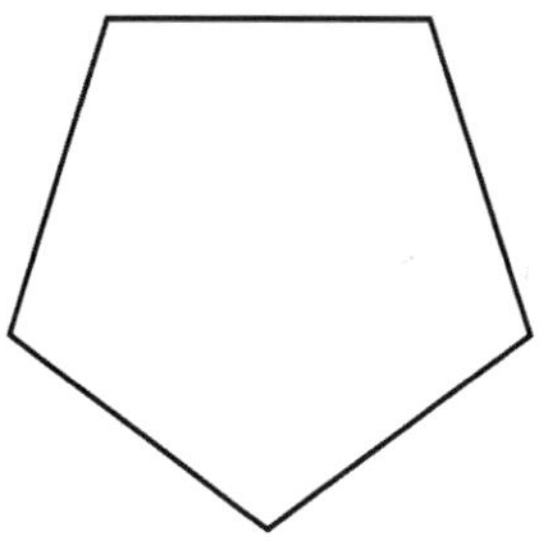

4. 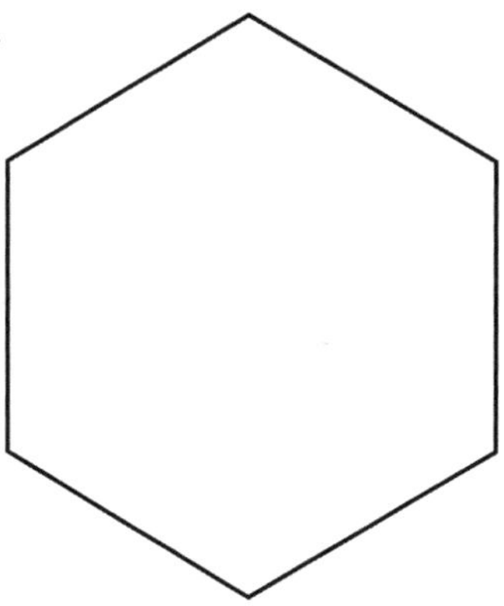

Name: ______________________________

Date: ______________________________

Write the fraction of a meter for the width of the beads entered.
Use 10 or 100 as the denominator.

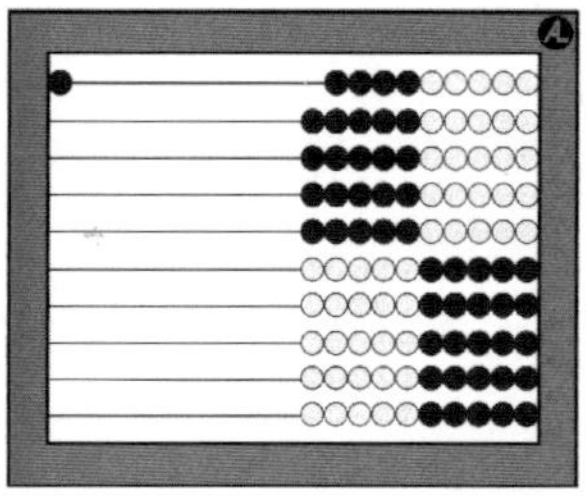
_____

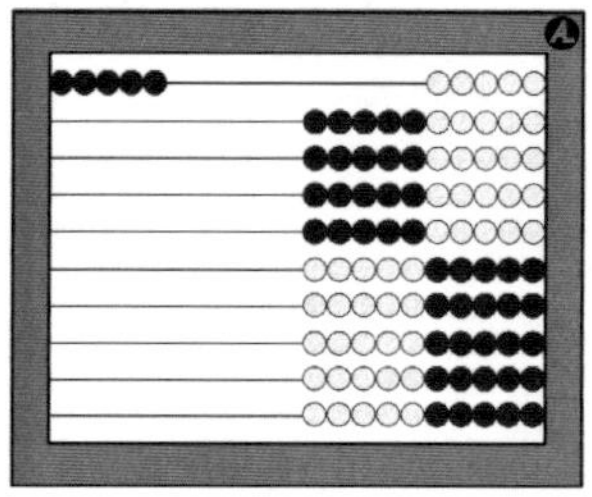
_____

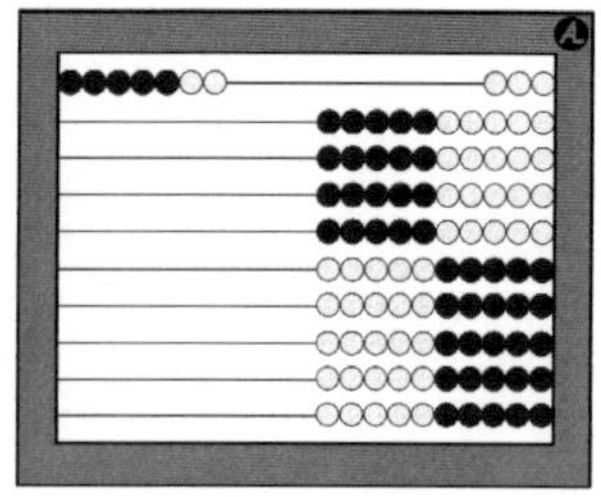
_____

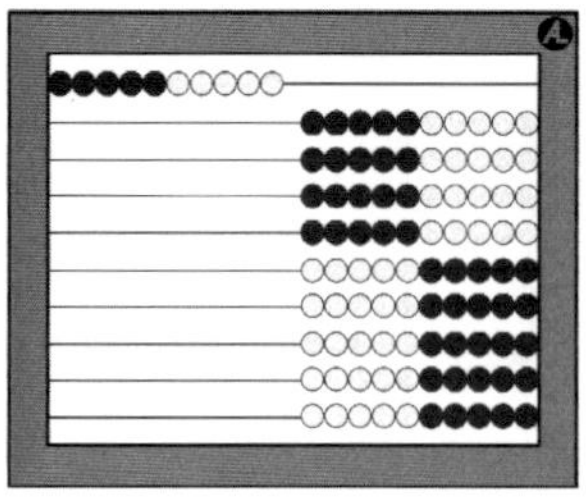
_____ _____

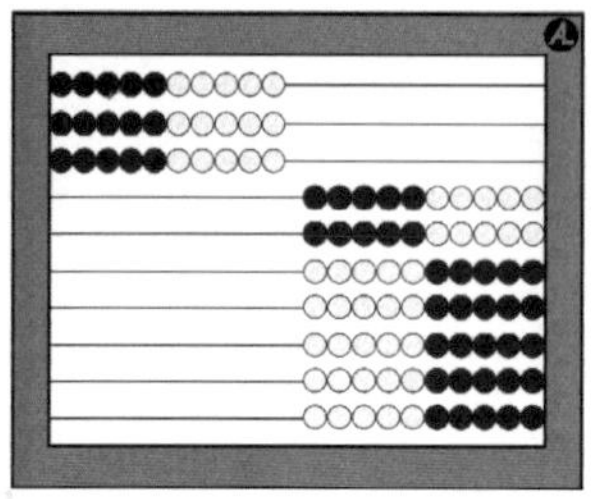
_____ _____

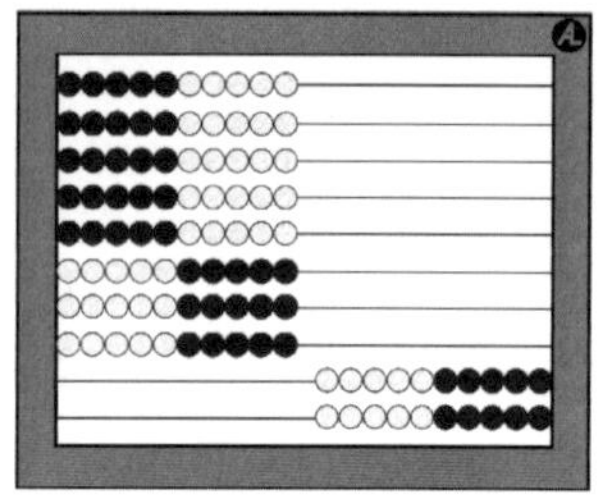
_____ _____

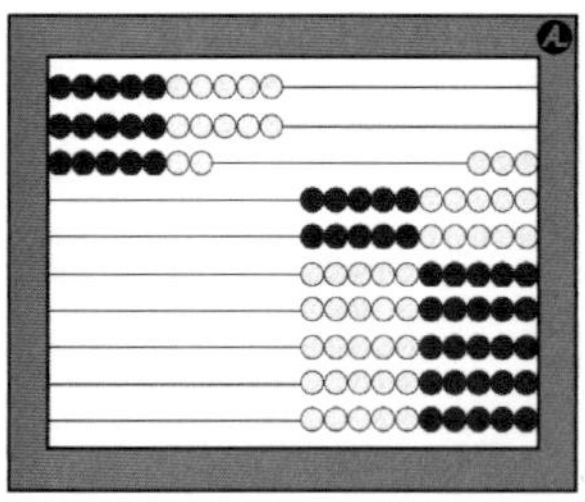
_____

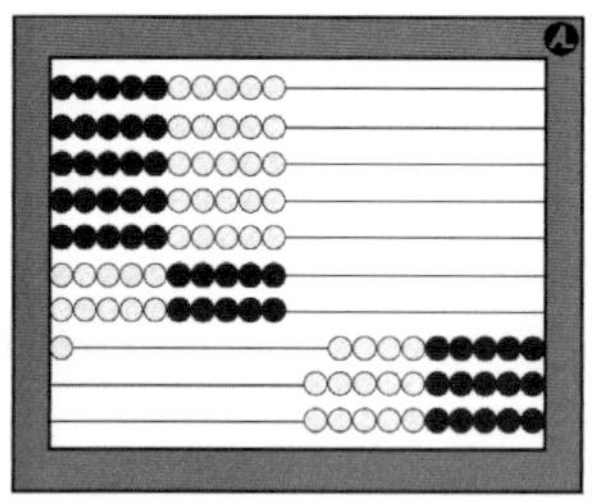
_____

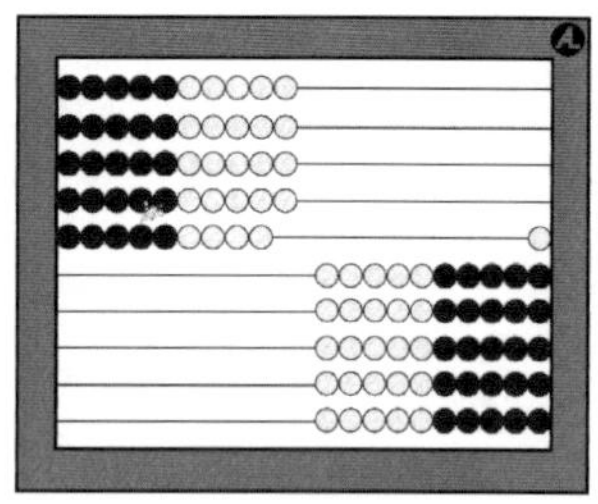
_____

Add the following fractions of a meter. Use your abacus to help.
Use 10 or 100 as the denominator.

$\frac{2}{100} + \frac{6}{100} =$ _____ $\frac{7}{100} + \frac{9}{100} =$ _____ $\frac{43}{100} + \frac{8}{100} =$ _____

$\frac{4}{10} + \frac{3}{10} =$ _____ $\frac{5}{10} + \frac{4}{10} =$ _____ $\frac{6}{10} + \frac{3}{10} =$ _____

$\frac{8}{10} + \frac{6}{100} =$ _____ $\frac{5}{10} + \frac{3}{100} =$ _____ $\frac{7}{10} + \frac{16}{100} =$ _____

$\frac{4}{10} + \frac{58}{100} =$ _____ $\frac{33}{100} + \frac{3}{10} =$ _____ $\frac{87}{100} + \frac{1}{10} =$ _____

Name: ______________________

Date: ______________________

Write the equation to show the action on the abacus.

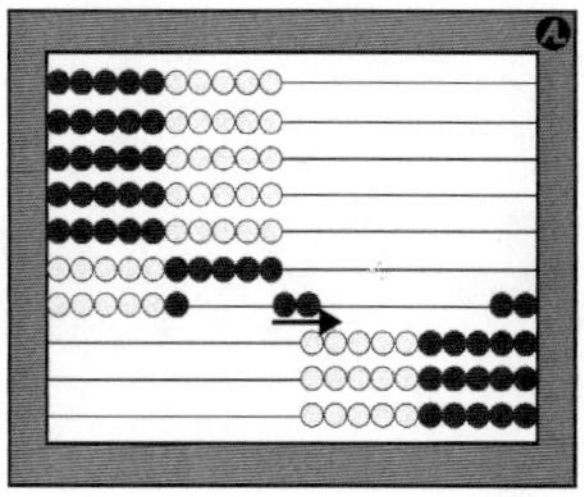

____

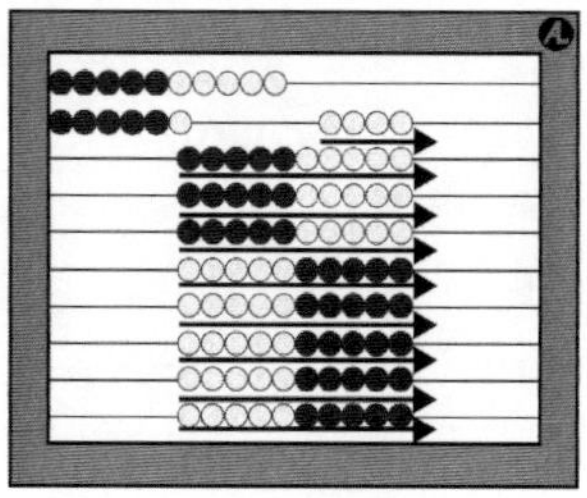

____

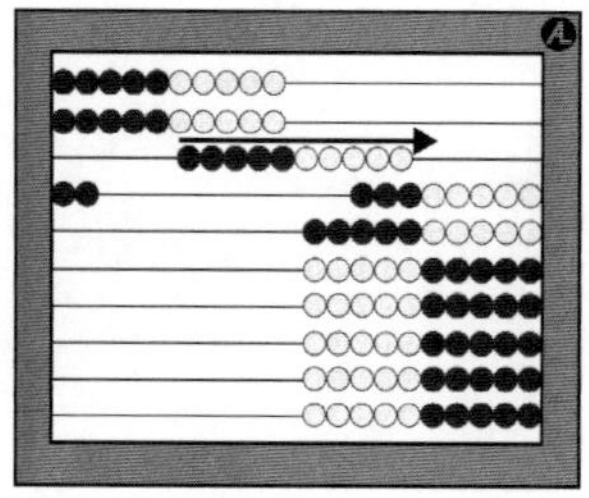

____

Perform the operation shown on the following fractions. Use your abacus to help. Use 10 or 100 as the denominators.

$\frac{9}{100} - \frac{6}{100} =$ ______ $\frac{11}{100} - \frac{1}{100} =$ ______ $\frac{43}{100} - \frac{3}{100} =$ ______

$\frac{6}{10} - \frac{3}{10} =$ ______ $1 - \frac{4}{10} =$ ______ $1 - \frac{83}{100} =$ ______

$\frac{2}{10} \times 3 =$ ______ $\frac{5}{10} - \frac{3}{100} =$ ______ $\frac{7}{10} - \frac{16}{100} =$ ______

$\frac{4}{10} - \frac{8}{100} =$ ______ $\frac{33}{100} - \frac{3}{10} =$ ______ $\frac{87}{100} - \frac{1}{10} =$ ______

Write <, =, or > on the lines.

$\frac{9}{100}$ ______ $\frac{6}{100}$ $\frac{9}{100} + \frac{4}{10}$ ______ $\frac{13}{100}$

$\frac{40}{100}$ ______ $\frac{4}{10}$ $\frac{81}{100}$ ______ $\frac{8}{10} + \frac{1}{10}$

$\frac{7}{10}$ ______ $\frac{87}{100}$ $\frac{4}{10}$ ______ $\frac{50}{100} - \frac{1}{10}$

$\frac{19}{100}$ ______ $\frac{2}{10}$ $1 - \frac{37}{100}$ ______ $\frac{6}{10}$

$\frac{3}{10}$ ______ $\frac{23}{100}$ $\frac{1}{10}$ ______ $\frac{5}{100} \times 2$

Worksheet 49, Decimal Fractions on the Vertical Abacus

Name: ______________________

Date: ______________________

Add the whole numbers. Use your abacus turned clockwise 90°.
Keep 10 or 100 as the denominators for the fractions.

$\begin{array}{r} 23 \\ +\,36 \\ \hline \end{array}$	$\begin{array}{r} 2\frac{3}{10} \\ +\,3\frac{6}{10} \\ \hline \end{array}$	$\begin{array}{r} 41 \\ +\,23 \\ \hline \end{array}$	$\begin{array}{r} 4\frac{1}{10} \\ +\,2\frac{3}{10} \\ \hline \end{array}$
$\begin{array}{r} 14 \\ +\,27 \\ \hline \end{array}$	$\begin{array}{r} 1\frac{4}{10} \\ +\,2\frac{7}{10} \\ \hline \end{array}$	$\begin{array}{r} 39 \\ +\,53 \\ \hline \end{array}$	$\begin{array}{r} 3\frac{9}{10} \\ +\,5\frac{3}{10} \\ \hline \end{array}$
$\begin{array}{r} 3\frac{9}{10} \\ +\,2\frac{8}{10} \\ \hline \end{array}$	$\begin{array}{r} \frac{39}{100} \\ +\,\frac{28}{100} \\ \hline \end{array}$	$\begin{array}{r} 6\frac{8}{10} \\ +\,9\frac{7}{10} \\ \hline \end{array}$	$\begin{array}{r} \frac{68}{100} \\ +\,\frac{97}{100} \\ \hline \end{array}$

Write your scores for the Corners™ with Tenths game.

Name: ____________________

Date: ____________________

Write the quantities shown using fractions and decimal points.

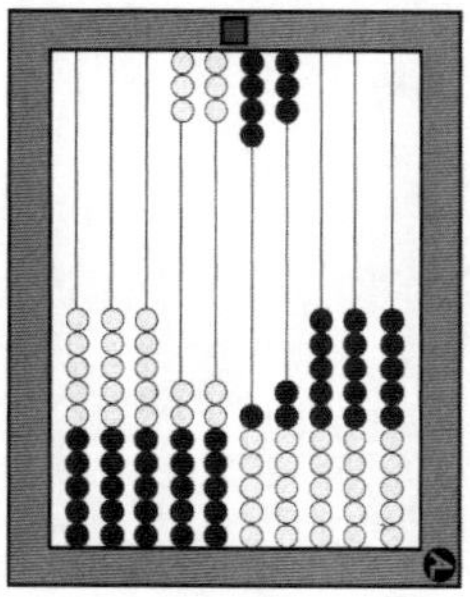

______ ______

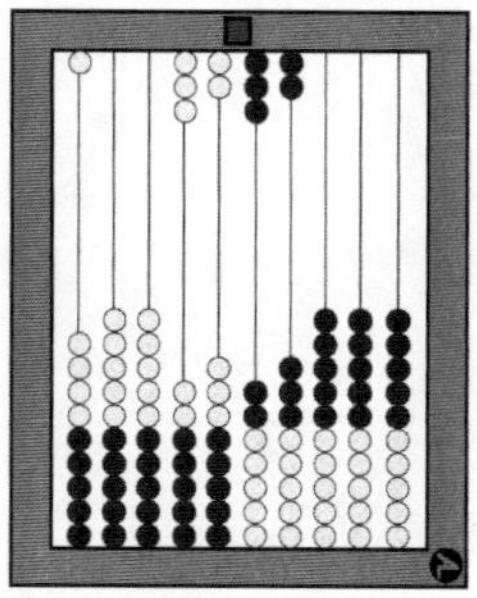

______ ______

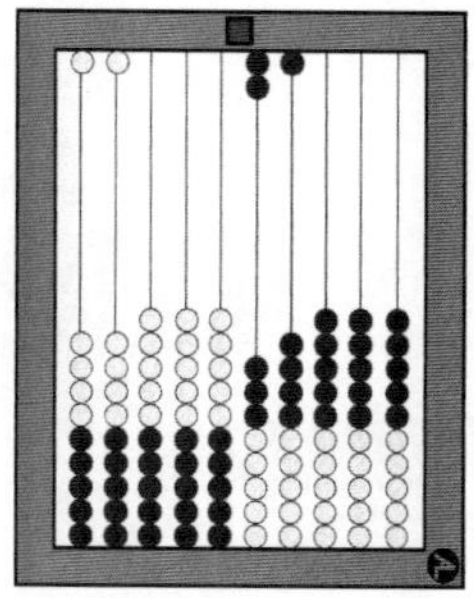

______ ______

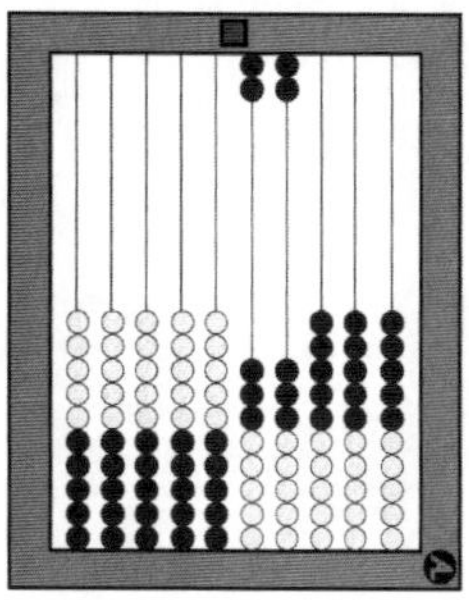

______ ______

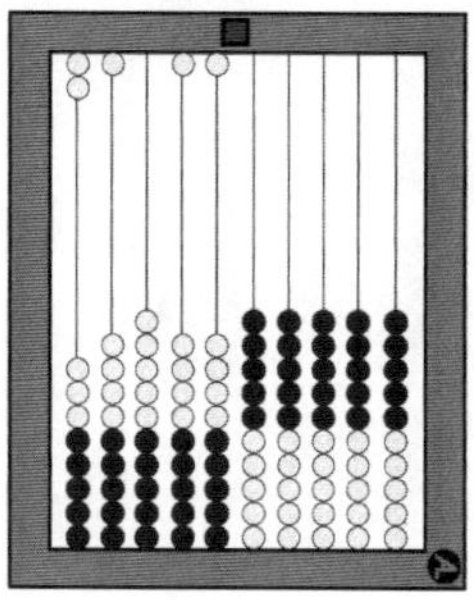

______ ______

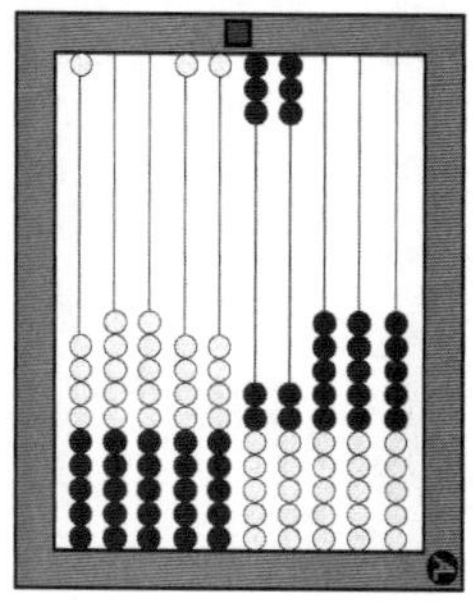

______ ______

Add the following.

$$\begin{array}{r} 2\frac{5}{10} \\ +\,5\frac{4}{10} \\ \hline \end{array} \qquad \begin{array}{r} 2.5 \\ +\,5.4 \\ \hline \end{array} \qquad \begin{array}{r} 3\frac{1}{10} \\ +\,8\frac{3}{10} \\ \hline \end{array} \qquad \begin{array}{r} 3.1 \\ +\,8.3 \\ \hline \end{array}$$

$$\begin{array}{r} 5\frac{6}{10} \\ +\,2\frac{9}{10} \\ \hline \end{array} \qquad \begin{array}{r} 5.6 \\ +\,2.9 \\ \hline \end{array} \qquad \begin{array}{r} 7\frac{5}{10} \\ +\,8\frac{8}{10} \\ \hline \end{array} \qquad \begin{array}{r} 7.5 \\ +\,8.8 \\ \hline \end{array}$$

$$\begin{array}{r} 1\frac{15}{10} \\ +\,2\frac{9}{10} \\ \hline \end{array} \qquad \begin{array}{r} 2.5 \\ +\,2.9 \\ \hline \end{array} \qquad \begin{array}{r} 3\frac{4}{10} \\ +\,\frac{12}{10} \\ \hline \end{array} \qquad \begin{array}{r} 3.4 \\ +\,1.2 \\ \hline \end{array}$$

Name: ______________________________

Date: ______________________________

Write the quantities shown using fractions and decimal points.

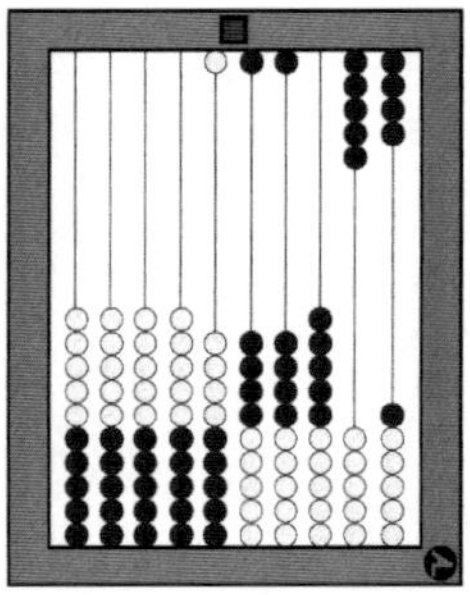

________ ________

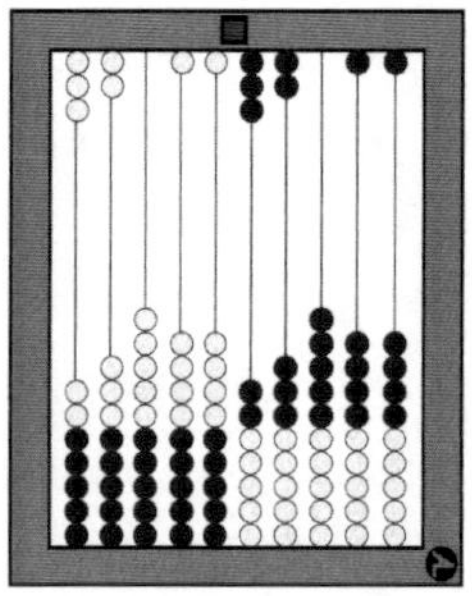

________ ________

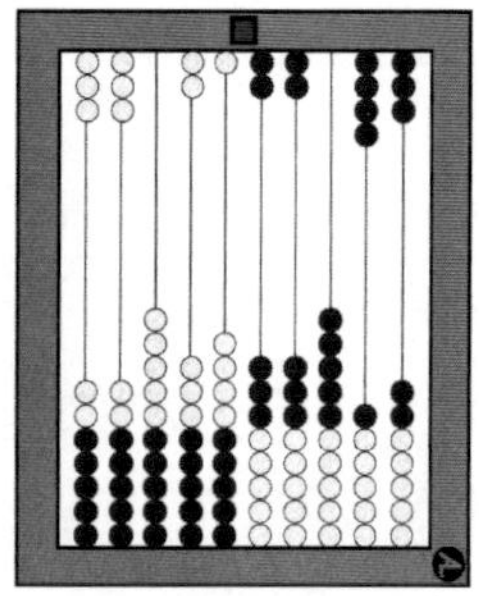

________ ________

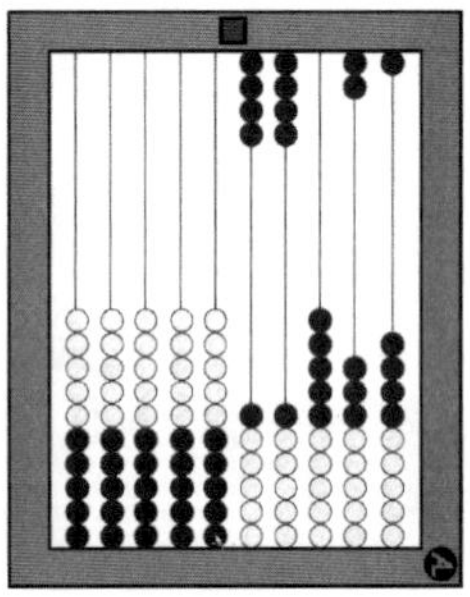

________ ________

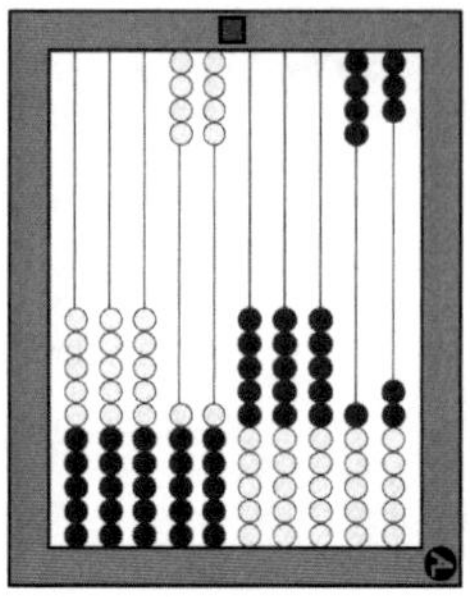

________ ________

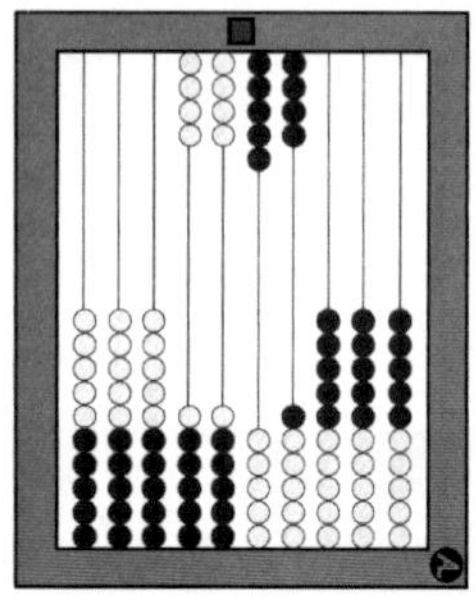

________ ________

Subtract the following.

21.6 − 3.5	9.3 − 5.6	10.0 − 8.5	9.1 − 8.3
11.63 − 2.31	9.47 − 2.87	9.53 − 5.28	7.41 − 5.53
5.68 − 2.08	5.15 − 2.90	8.00 − 1.25	3.40 − 1.25

Worksheet 52, Decimal and Fraction Practice

Name: ______________________________

Date: ______________________________

Match the decimals with the fractions and with the words.

$\frac{4}{100}$	4.4	four hundred
$\frac{4}{1000}$	40.4	four hundredths
$4\frac{4}{10}$	400	four and four tenths
$40\frac{4}{10}$	0.4	four tenths
$44\frac{4}{10}$	.04	forty and four tenths
400	.004	forty-four and four tenths
$\frac{4}{10}$	44.4	four thousandths
$4\frac{4}{100}$	4.04	four and four hundredths

Write these fractions as decimals.

$\frac{4}{100}$ ________

$\frac{17}{100}$ ________

$3\frac{2}{10}$ ________

$27\frac{8}{10}$ ________

$9\frac{50}{100}$ ________

$3\frac{10}{10}$ ________

$\frac{9}{10}$ ________

$1\frac{7}{100}$ ________

Write these decimals as fractions.

0.7 ________

.7 ________

0.95 ________

8.15 ________

.06 ________

.82 ________

3.20 ________

.003 ________

Put these fractions and decimals in order from least to greatest.

$\frac{5}{10}$	$\frac{5}{100}$	.04	______________
$\frac{300}{100}$	.03	.33	______________
.5	$\frac{49}{100}$	$\frac{51}{100}$	______________
1.0	$\frac{9}{10}$	$\frac{1}{3}$	______________
.06	$\frac{3}{4}$	$\frac{50}{100}$	______________

Name: ______________________

Date: ______________________

Fill in the chart below.

Coins	Value in ¢	Number in a Dollar	Fraction of a Dollar Fraction	Hundredths	Decimal
Penny					
Nickel					
Dime					
Quarter					

Amounts on checks are usually written two ways: in numerals using hundredths of a dollar and in words for dollars and hundredths of a fraction for cents. Write the following amounts as for a check.

84 dollars and 4 dimes	$84.40
Eighty-four and 40/100	Dollars
11 dollars and 7 dimes	
	Dollars
172 dollars and 1 nickel	
	Dollars
49 dollars and 20 pennies	
	Dollars
8 dollars and 2 quarters	
	Dollars
15 dollars and 5 quarters	
	Dollars
18 dollars and 20 nickels	
	Dollars
21 dollars, 10 dimes, and 4 pennies	
	Dollars
54 dollars, 2 dimes, and 40 nickels	
	Dollars
34 dollars, 6 quarters, and 10 pennies	
	Dollars

Worksheet 54–A, Review 5

Name: ______________________________

Date: ______________________________

Write only the answers.

____

____

____

Write the answers.

$2 - \frac{9}{10} =$ ______

$\frac{63}{100} +$ ______ $= 1$

$3\frac{5}{10} - 1.5 =$ ______

Do the arithmetic. Use check numbers.

4679 + 13,674   8387 – 7793   2839 × 7   4815 ÷ 3

Solve the problem.

A turtle walked around the path shown.
How many meters did it walk?

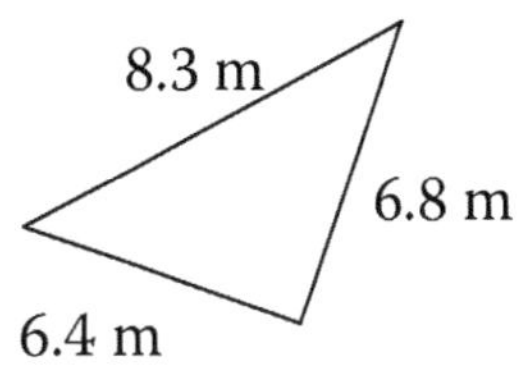

Put these fractions in order from least to greatest.

$\frac{7}{10}$ $\frac{1}{100}$ .6 $\frac{1}{2}$ $\frac{24}{8}$ $\frac{7}{7}$

____

Circle the numbers that are greater than one.

1.96 $\frac{63}{100}$ $\frac{2}{3}$ $\frac{6}{6}$ $\frac{12}{10}$ $\frac{111}{100}$

Circle the prime number in each row.

27   25   49   61

17   58   93   40

87   88   89   85

Write >, <, or = on the lines.

0.46 ____ .46

$\frac{2}{7}$ ____ $\frac{4}{7}$

$1 + \frac{7}{10}$ ____ $\frac{163}{100}$

$1 + \frac{1}{100}$ ____ 1.01

$\frac{7}{10}$ ____ $\frac{7}{8}$

$\frac{1}{100} + \frac{7}{10}$ ____ .71

5 dimes ____ \$0.60

0.86 ____ $\frac{8}{10}$

Name: ______________________

Date: ______________________

Write only the answers.

____

____

____

Write the answers.

$2 - \frac{7}{10} =$ ______

$\frac{56}{100} +$ ______ $= 1$

$4\frac{3}{10} - 1.3 =$ ______

Do the arithmetic. Use check numbers.

7956 + 21,786   9674 – 8685   7594 × 8   5624 ÷ 4

Solve the problem.

A tortoise walked around the rectangular path shown. How many meters did it walk?

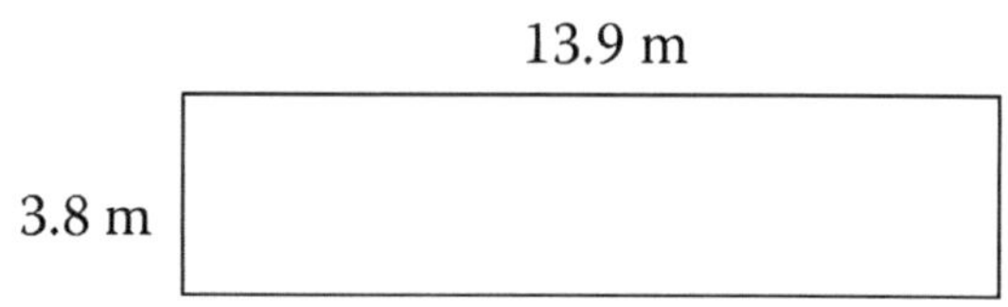

Put these fractions in order from least to greatest.

$\frac{1}{10}$ $\frac{100}{100}$ .6 $\frac{1}{4}$ $\frac{20}{4}$ $\frac{1}{7}$

____

Write >, <, or = on the lines.

.27 ____ 0.27

$\frac{3}{9}$ ____ $\frac{2}{9}$

$1 + \frac{2}{10}$ ____ $\frac{123}{100}$

$2 + \frac{2}{100}$ ____ 2.02

$\frac{6}{12}$ ____ $\frac{6}{8}$

$\frac{5}{100} + \frac{8}{10}$ ____ .85

5 nickels ____ \$0.05

0.94 ____ $\frac{9}{10}$

Circle the numbers that are greater than one.

0.99 $\frac{103}{100}$ $\frac{4}{3}$ $\frac{8}{8}$ $\frac{21}{10}$ $\frac{74}{100}$

Circle the prime number in each row.

21	23	39	15
7	18	99	36
49	45	81	79

Name: ________________________

Date: ________________________

Use your calculator to find the answers.

$6 + 17 \times 8 =$ ____ $34 \times 8 + 52 \times 4 =$ ____ $34 \times 78 - 773 \times 3 =$ ____
$(17 + 35) \times 8 =$ ____ $7 \times 6 + 7 \times 3 + 7 \times 11 =$ ____ $7 \times (6 + 3 + 11) =$ ____
$(3 + 5) \times (4 + 7) =$ ____ $3 + 5 \times (4 + 7) =$ ____ $3 + 5 \times 4 + 7 =$ ____ $(3 + 5) \times 4 + 7 =$ ____

$(8 + 6) \times (2 + 11) =$ ____

$4 \times (14 + 6) =$ ____

$12 \times 6 + 12 \times 3 + 12 \times 3 =$ ____

$12 \times (6 + 3 + 3) =$ ____

$21 + 42 \times (39 + 1) =$ ____

$(21 + 42) \times (39 + 1) =$ ____

$21 + 42 \times 39 + 1 =$ ____

$(21 + 42) \times 39 + 1 =$ ____

Write an equation for the problem and solve it.

Mr. Mason bought 1 item for \$10, 7 items for \$4 each, and 12 items for \$3 each. What was the total bill?

____

____

Write your own story problem for this equation. Then solve the equation.

$n = 5 \times 20 + 2 \times 30$

____

____

____

____

____

____

Name: ______________________________

Date: ______________________________

1. Use your calculator to find the answers. Write the answers as the calculator shows and as money in dollars and cents.

	3 dimes	4 nickels	2 quarters	5 quarters	8 quarters	4 pennies
Calculator						
Dollars	$0.30					

Solve these problems using a calculator. Write the equations.

2. Carson bought a dozen apples for $3.60. What did each apple cost?

3. A group of eight people paid $20.80 for pizza. What was the cost per person?

4. For a party Meggie bought 18 sets of cards that cost 59¢ each. She also bought 18 small favors that cost 16¢ each. What was the total cost?

5. Jon and Kenzie went shopping for a gift. They spent nine dollars and 98 cents for a book and one dollar and fifty-five cents for wrapping paper and a quarter for a bow. What is the total cost for the gift? They each paid half. What did they each pay?

6. Matthew has five dollars to buy parts. He bought 6 nuts at 20¢ each, 5 bolts at 25¢ each, and 9 washers at 5¢ each. How much money does he have left?

7. Anna bought one item that cost $4, two items that are 89¢ each, and eight items that cost 49¢. Anna brought $10.50 to the store. How much money did Anna have left?

Worksheet 57, Decimals on a Number Line

Name: ______________________________

Date: ______________________________

1. Write the place on the number line, shown by a letter, by the corresponding number.

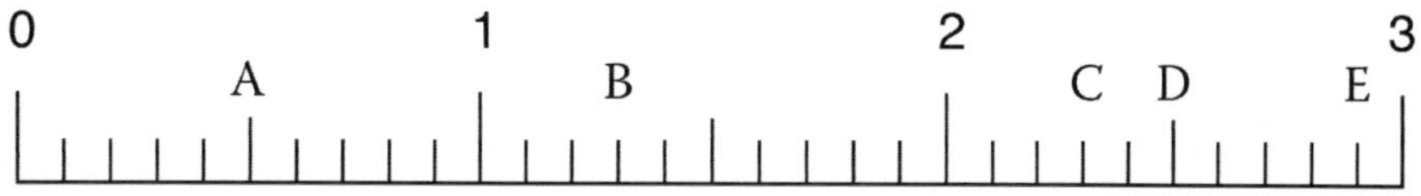

$\frac{1}{2}$ ______ $2\frac{1}{2}$ ______ 2.9 ______ 1.3 ______ 2.3 ______ 0.5 ______

2. Write the letter on the line at the correct place: T is 0.30, U is 0.16, V is 0.09, W is 0.26, X is 0.01, Y is 0.12, and Z is 0.21.

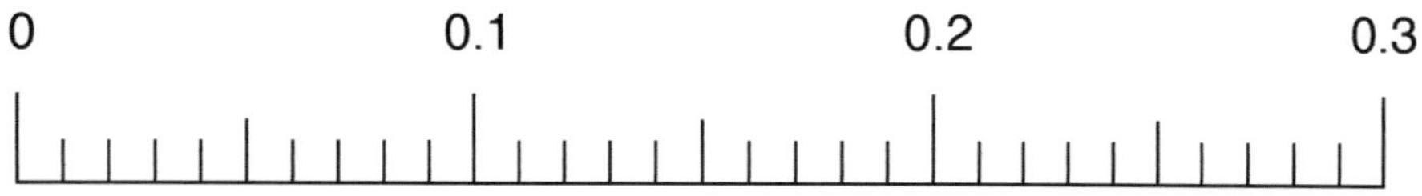

___ ___ ___ ___ ___ ___ ___

3. Write the letter on the corresponding line.

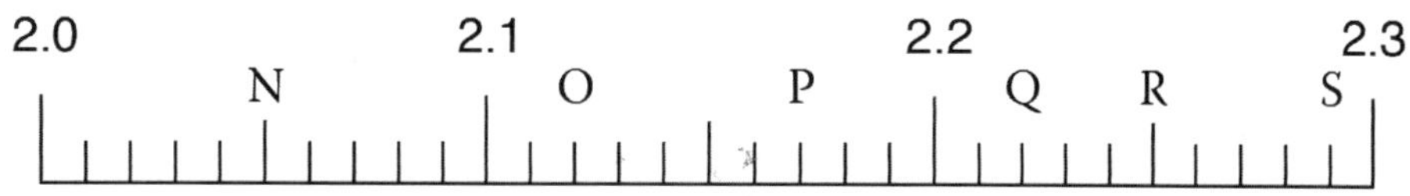

2.25 _____ 2.05 _____ 2.12 _____ 2.29 _____ 2.22 _____ 2.17 _____

4. Write the letter on the line at the correct place: I is 3.16, M is 3.22, C is 3.10, E is 3.08, D is 3.05, S is 3.30, L is 3.27, and A is 3.25.

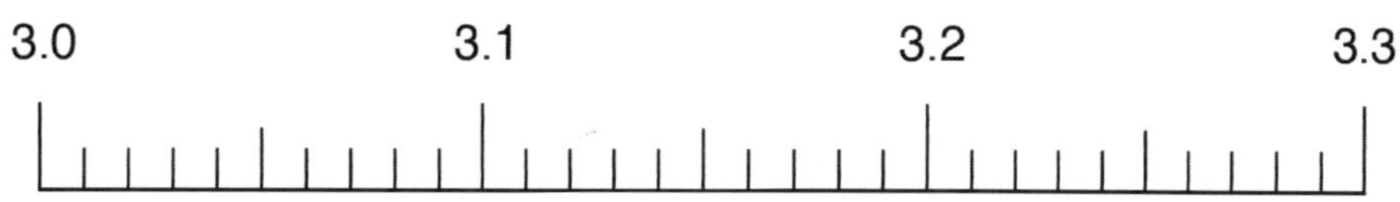

___ ___ ___ ___ ___ ___ ___ ___

Worksheet 58, Measuring in Tenths of an Inch and a Mile

Name: ______________________________

Date: ______________________________

Measure the figures in tenths of an inch and find the perimeters.
Write your solutions inside the figures.

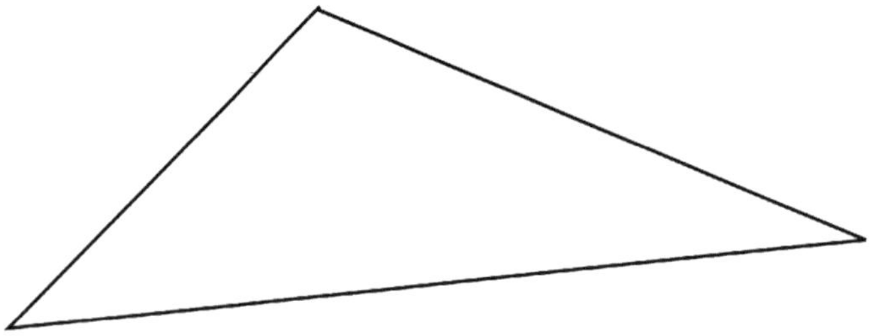

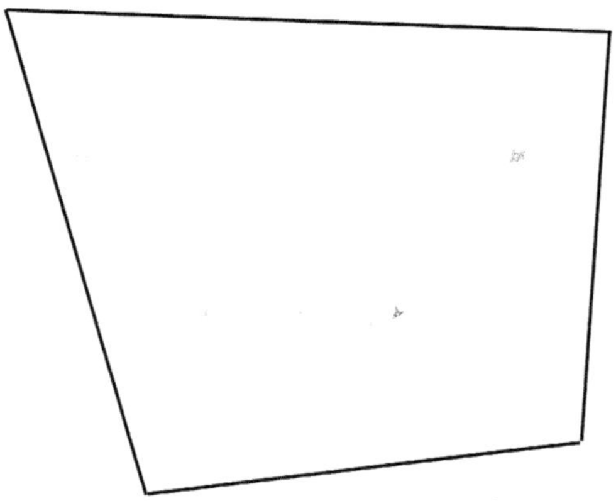

5.8 in. + 2.7 in. = ________

5.5 in. + 7.9 in. = ________

8.2 in. – 2.6 in. = ________

2.5 in. – 0.6 in. = ________

9.5 in. + ________ = 1 ft

10.8 in. + ________ = 1 ft.

0.8 in. + 5.3 in. = ________

4.3 in. + 7 in. = ________

8.4 in. – 2.7 in. = ________

6.7 in. – 5.5 in. = ________

2.2 in. + ________ = 1 ft

7.4 in. + ________ = 1 ft

Name: ______________________________

Date: ______________________________

1. Draw an arrow to show 1 decimeter + 1 centimeter + 1 millimeter.

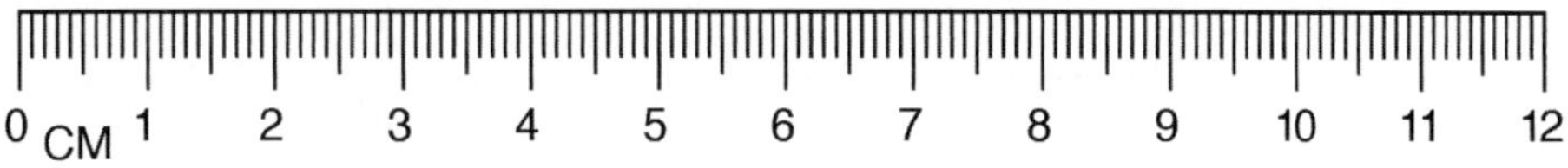

How many millimeters is 1 dm + 1 cm + 1 mm? __________

Write it as a decimal fraction of a centimeter. __________

Write it as a decimal fraction of a meter. __________

2. Write these measurements as decimal fractions of a meter and then add them.

	Decimal fraction of a meter	Sum
1 dm + 2 cm	0.1 + 0.02	0.12 m
1 dm + 4 mm		
2 dm + 16 mm		
1 m + 2 dm		
4 m + 3 cm		
2 m + 3 dm + 12 cm		
3 m + 7 dm + 7 mm		
2 m + 5 cm + 1 mm		
5 m + 2 mm		

3. Measure the sides of the large and small triangles. Use centimeters, millimeters, and meters in decimal fractions to fill in the table.

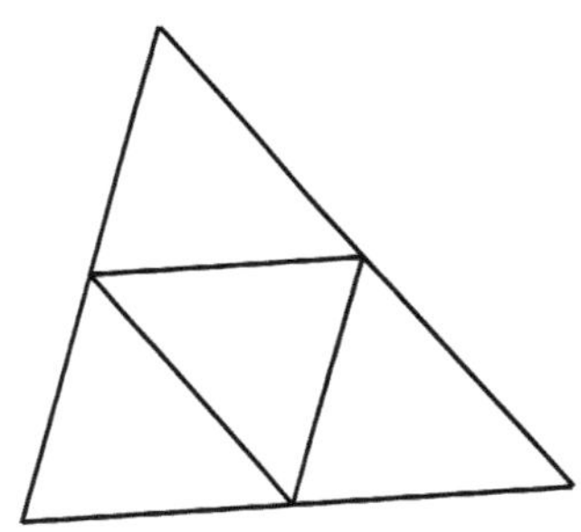

Large Triangle			Small Triangles		
cm	mm	m	cm	mm	m

Name: ______________________________

Date: ______________________________

Warm-Up

Round the following amounts by putting them into the column they are closest to:

$123.23 $124.70 $124.24 $125.67
$124.00 $124.91 $123.99 $126.03
$125.99 $122.50 $122.90 $124.61
$125.46 $123.47 $123.56 $125.72

$123	$124	$125	$126

These are signs at two gas stations At these stations, the amount that customers pay is rounded up only if the fraction of a cent is 5 or more.

UNLEADED	$3.48\frac{9}{10}$
PREMIUM	$3.99\frac{9}{10}$
DIESEL	$4.09\frac{9}{10}$

Jack's Station

UNLEADED	3.49^9
PREMIUM	3.99^9
DIESEL	4.05^9

Jill's Station

1. What does the $\frac{9}{10}$ mean at Jack's Station? ______________________________
2. What does the little 9 mean at Jill's Station? ______________________________
3. Which price is the same at the two stations? ______________________________

Fill in the charts, giving the cost of the fuel in dollars and cents. Use your calculators.

Number of Gallons	Unleaded Gasoline	
	Jack's	Jill's
1	$3.49	$3.50
5		
10		
20		

Number of Gallons	Diesel	
	Jack's	Jill's
30		
100		
200		
300		

4. How much will you save by buying 20 gallons of unleaded at Jack's rather than Jill's? ______
5. What is the cost of 15 gallons of premium gasoline? _________
6. What would be the cost of 15 gallons of premium gasoline at $4.00 a gallon? _________

Name: ______________________________

Date: ______________________________

Write only the answers.

____

____

____

Write the answers.

$2 – 50¢ = ______

1.69 + ______ = 2

$2\frac{5}{10} \times 2 =$ ______

Do the calculations.

3542 – 1576   9056 + 37 + 249   2835 ÷ 7   49 + 49 + 49 + 49

Solve the problem.

UNLEADED	2.49^9
PREMIUM	2.69^9
DIESEL	3.05^9

Robin bought 10 gallons of premium gasoline. How much was the change from a hundred dollar bill?

Fill in the tables below.

m	dm	cm	mm
		100	
0.5			
	9		

1 yard (yd) is 3 feet (ft). Use fractions.

yd	ft	in.
1		
		18
	2	

Use your ruler and find the perimeter of the triangle in centimeters. Also find the perimeter in millimeters.

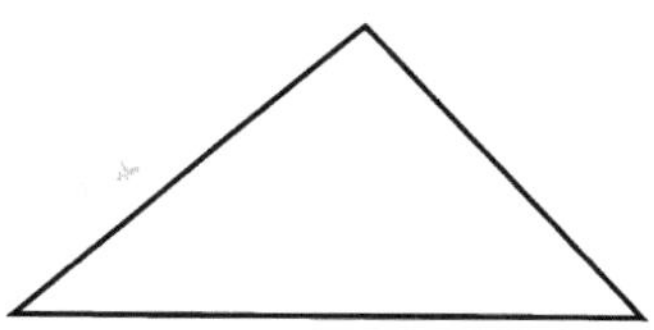

Name: ______________________________

Date: ______________________________

Write only the answers.

____

____

____

Write the answers.

$2.50 – $1 = ______

1.57 + ______ = 2

$1\frac{5}{10} \times 4 =$ ______

Do the calculations.

4786 – 3797

8704 + 68 + 457

5624 ÷ 8

59 + 59 + 59 + 59

Solve the problem.

UNLEADED	2.99^9
PREMIUM	3.19^9
DIESEL	3.05^9

Alex bought 10 gallons of unleaded gasoline. How much was the change from a hundred dollar bill?

Fill in the tables below.

m	dm	cm	mm
		100	
1.5			
	1		

1 yard (yd) is 3 feet (ft). Use fractions.

yd	ft	in.
1		
	$1\frac{1}{2}$	
		12

Use your ruler and find the perimeter of the triangle in centimeters. Also find the perimeter in millimeters.

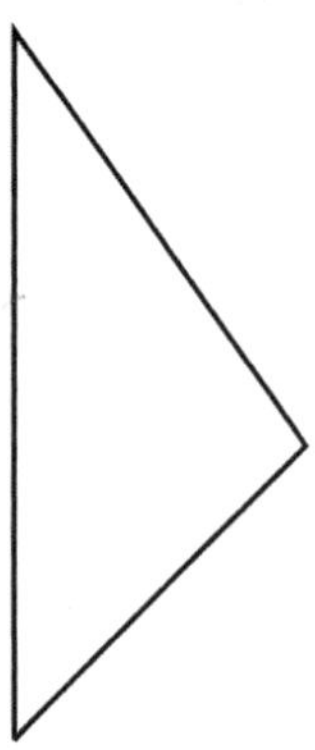

Name: ______________________________

Date: ______________________________

Circle the number of beads and write the percent.

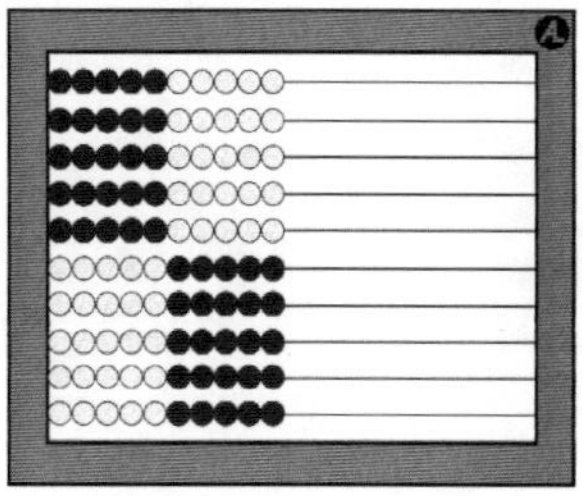

Circle $\frac{1}{2}$ of the beads.

Percent: __________

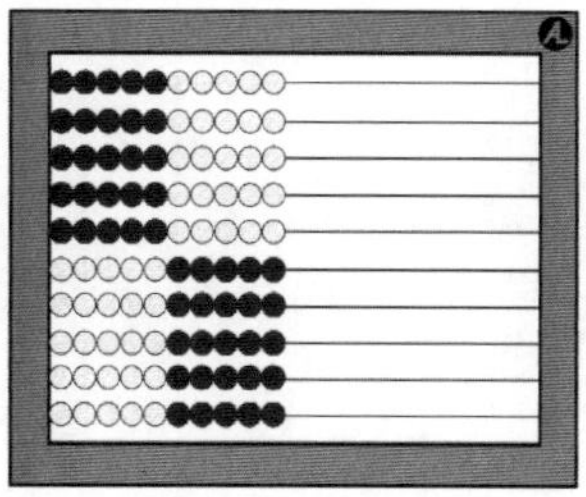

Circle $\frac{1}{3}$ of the beads.

Percent: __________

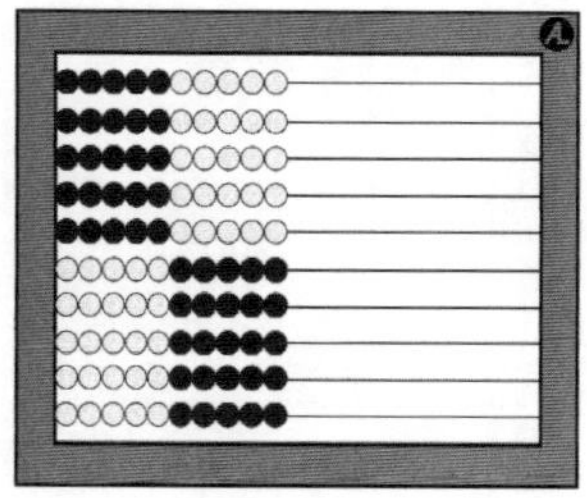

Circle $\frac{1}{8}$ of the beads.

Percent: __________

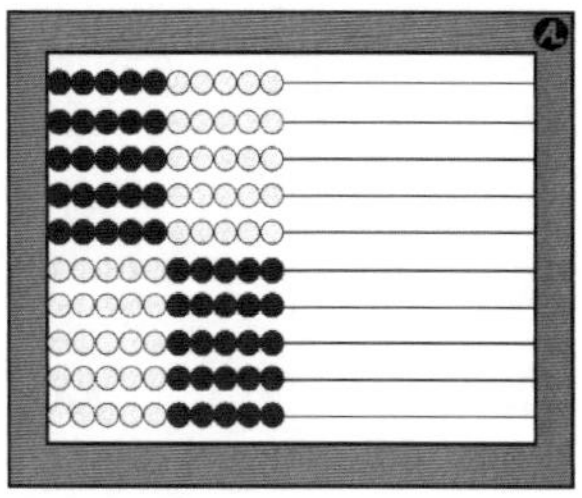

Circle $\frac{3}{5}$ of the beads.

Percent: __________

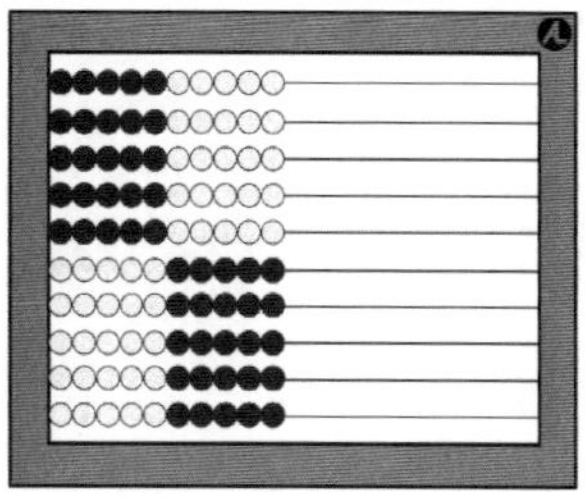

Circle $\frac{3}{4}$ of the beads.

Percent: __________

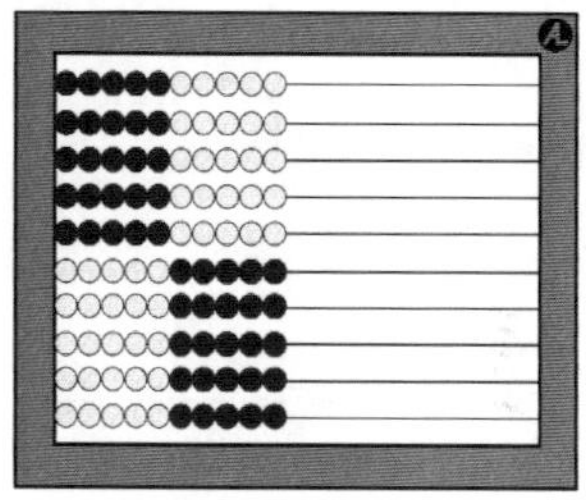

Circle $\frac{9}{10}$ of the beads.

Percent: __________

Fill in the chart below. You will need it for a future lesson.

	Write the percent.		Write the percent.		Write the percent.		Write the percent.
1		$\frac{1}{2}$					
$\frac{1}{3}$		$\frac{2}{3}$					
$\frac{1}{4}$		$\frac{3}{4}$					
$\frac{1}{5}$		$\frac{2}{5}$		$\frac{3}{5}$		$\frac{4}{5}$	
$\frac{1}{8}$		$\frac{3}{8}$		$\frac{5}{8}$		$\frac{7}{8}$	
$\frac{1}{10}$		$\frac{3}{10}$		$\frac{7}{10}$		$\frac{9}{10}$	

Name: ______________________________

Date: ______________________________

1. Crosshatch the following rectangles with the percents given.

50% 25% 100%

75% $33\frac{1}{3}\%$ $66\frac{2}{3}\%$

2. Put the letter by the corresponding fraction.

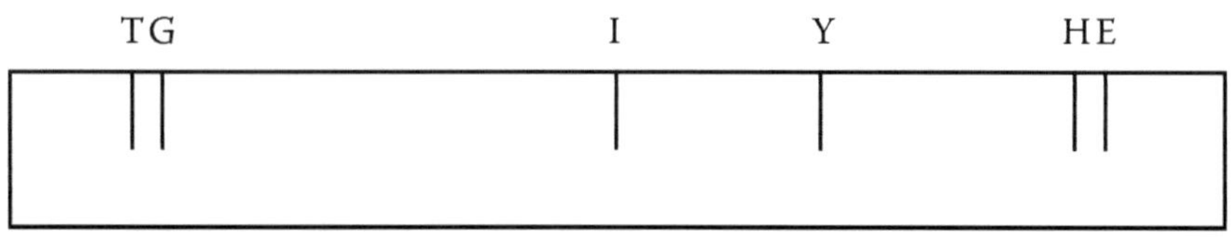

$\frac{9}{10}$ ____ $\frac{1}{2}$ ____ $\frac{1}{8}$ ____ $\frac{7}{8}$ ____ $\frac{1}{10}$ ____ $\frac{2}{3}$ ____

3. Put the letter by the corresponding fraction.

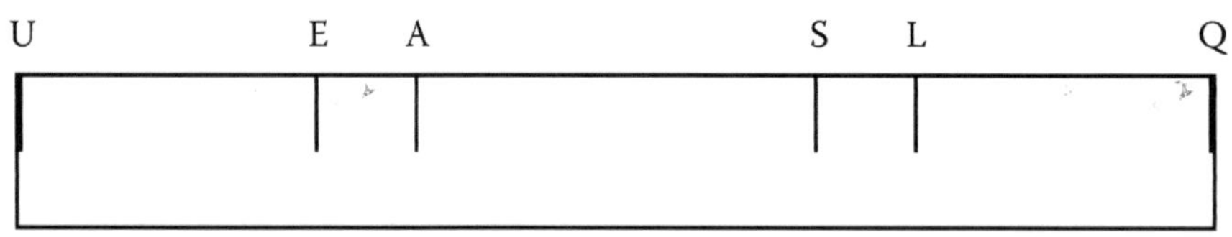

25% ____ 100% ____ 0% ____ $33\frac{1}{3}\%$ ____ 75% ____ $66\frac{2}{3}\%$ ____

4. Put the letter on the corresponding line.

U: $66\frac{2}{3}\%$ R: $\frac{1}{5}$ T: 90% P: $\frac{0}{3}$ D: 50% O: 25% C: $\frac{3}{4}$

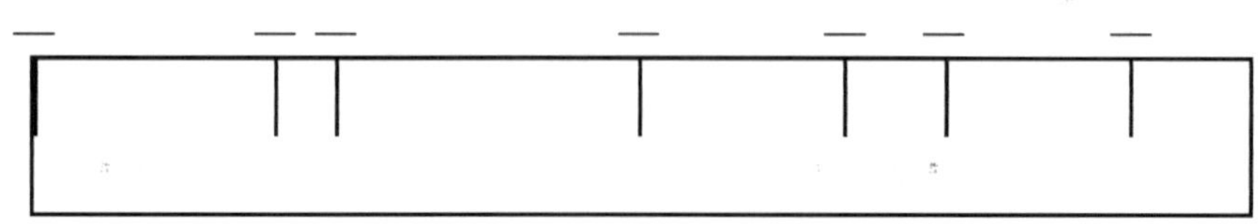

Name: ____________________

Date: ____________________

The questions refer to the numbers in the hundred chart.

1	2	3	4	5	6	7	8	9	10
11	12	13	14	15	16	17	18	19	20
21	22	23	24	25	26	27	28	29	30
31	32	33	34	35	36	37	38	39	40
41	42	43	44	45	46	47	48	49	50
51	52	53	54	55	56	57	58	59	60
61	62	63	64	65	66	67	68	69	70
71	72	73	74	75	76	77	78	79	80
81	82	83	84	85	86	87	88	89	90
91	92	93	94	95	96	97	98	99	100

1. What percentage of the numbers have only 1 digit? ____________
2. What percentage of the numbers have 3 digits? ____________
3. What percentage of the numbers have 2 digits? ____________
4. What percentage of the numbers are even numbers? ____________
5. What percentage of the numbers are odd numbers? ____________
6. What percentage of the numbers are multiples of 10? ____________
7. What percentage of the numbers are multiples of 5? ____________
8. What percentage of the multiples of 5 are even? ____________
9. What percentage of the numbers are not multiples of 5? ____________
10. What percentage of the numbers > 80? ____________
11. What percentage of the numbers < 15? ____________
12. What percentage of the numbers are perfect squares? ____________
    (Perfect squares are numbers such as 1, 4, 9, 16, that is,
    the products of 1 × 1, 2 × 2, 3 × 3 and so on.)
13. What percentage of the perfect squares are even? ____________
14. What percentage of the numbers are multiples of 9? ____________
15. What percentage of the multiples of 9 are multiples of 3? ____________
16. What percentage of the numbers in the first row are prime? ____________

Name: ____________________

Date: ____________________

The questions refer to the 50 states of the United States. Use your map and list of states.

1. What percentage of the states' names start with the letter *M*? __________
2. What percentage of the states' names start with the letter *N*? __________
3. What percentage of the names starting with *N* have two words? __________
4. What percentage of the states' names start with letters other than *M* or *N*? __________
5. What percentage of the states' names have two words? __________
6. What percentage of the states' names have only one word? __________
7. What percentage of the states were part of the 13 original colonies? __________
8. What percentage of the states joined after the original colonies? __________
9. What percentage of the states border the Mississippi River? __________
10. What percentage of the states have a coastline on the Pacific Ocean? __________
11. What percentage of the states border either the Atlantic Ocean or Gulf of Mexico? __________
12. What percentage of the states border Canada? __________
13. What percentage of the states border the Great Lakes? __________
14. What percentage border both a Great Lake and an ocean? __________
15. What percentage of the states border Mexico? __________
16. What percentage of the states border only other states? __________
17. What percentage of the states border Kentucky? __________
18. What percentage of the states do not border any other state? __________
19. What percentage of the states are in North America? __________
20. What percentage of the states are in South America? __________
21. What percentage of the states have borders that are only straight lines? __________
22. What percentage of the states have borders that are an arc of a circle? __________
(Hint: Look near Delaware.)

Name: ______________________________

Date: ______________________________

Solve the following problems.

1. Some shoes are on sale for 50% off. The original price (the price before the sale) is $48.30. What is the sale price?

2. The Reader's Bookstore is selling a book at 10% off. How much did the buyer save if the original price is $22? What is the sale price?

3. Lee can't decide which gift to buy. The red one is $5 and is 10% off. The green one is $6 and is 20% off. Which one is cheaper?

4. Prices have gone up 10% at the Little Store. What is the new price for cereal, which used to cost $4.00?

5. A jacket is on sale at a store for $35.88. There is a tag on it saying it was marked down 50%. What was the original price? Hint: first solve the problem with easier numbers.

Name: ______________________________

Date: ______________________________

Solve the following problems.

1. In a certain class 50% of the children are girls. There are 12 girls. How many children are in the class?

2. The usual tip at a restaurant is 15% of the cost of the food. Many people figure it out by first finding 10%, then finding 5%, which is half of 10%, and adding them together. What is the tip if the food costs $8.00?

3. What is the 15% tip if the food bill is $12.00? What is the total cost?

4. In some places people pay sales tax on certain things they buy. If the sales tax is 5%, what is the total bill for a car that cost $4000?

5. The original price for a game is $10.00. In Store A it went on sale at 10% off and then it went on sale again with 50% off of the sale price. In Store B it went on sale at 50% off and then it went on sale again with 10% off of the sale price. Which store has the better price?

Name: ______________________________

Date: ______________________________

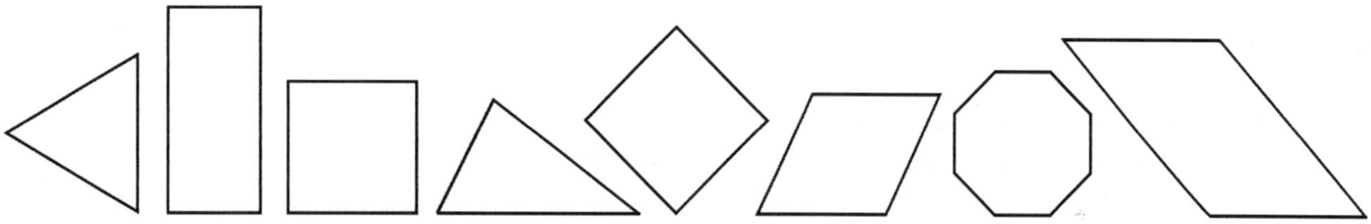

For each problem below, find the fraction of the polygons that are shown above. Use your fraction circle to show the answer, then record it below. Shade or crosshatch the circle so that it looks like the fraction circle answer.

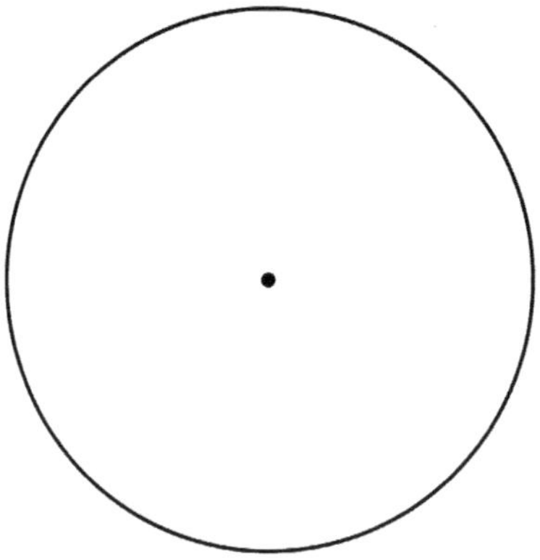

What fraction of the polygons have parallel sides? ____________

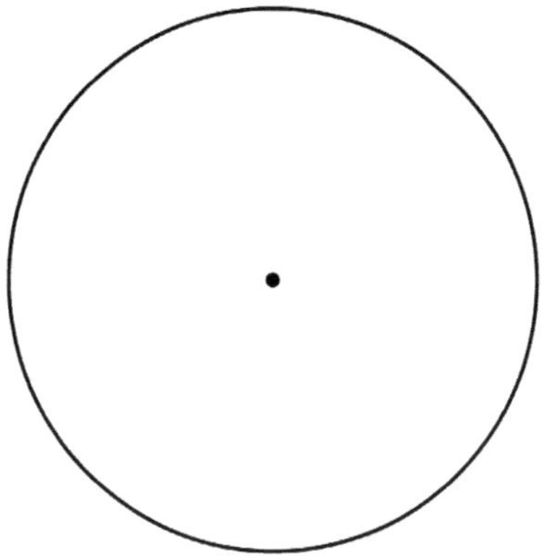

What fraction of the polygons have perpendicular lines? ____________

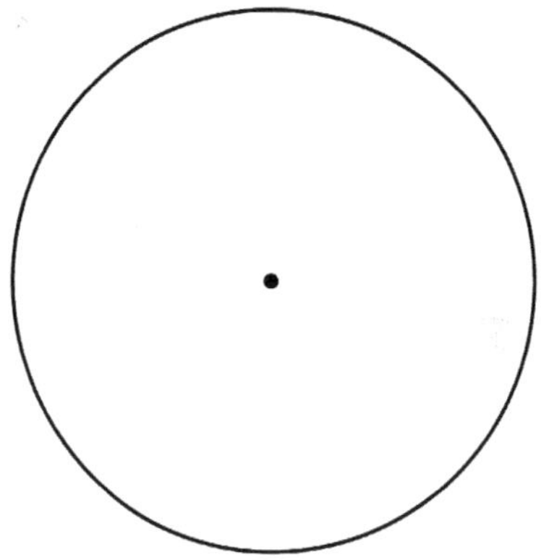

What fraction of the polygons are quadrilaterals? ____________

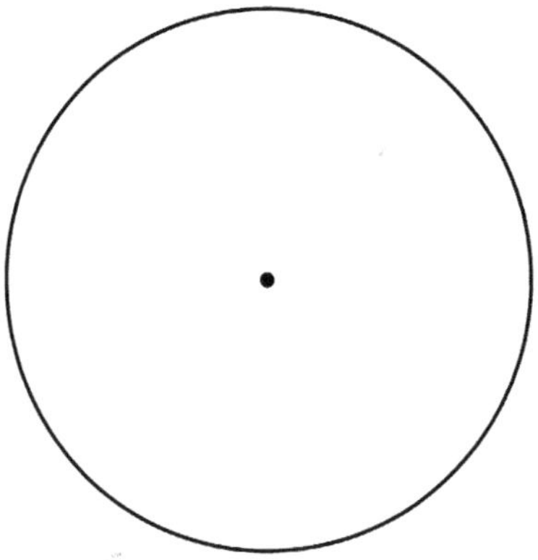

What fraction of the polygons are rectangles? ____________

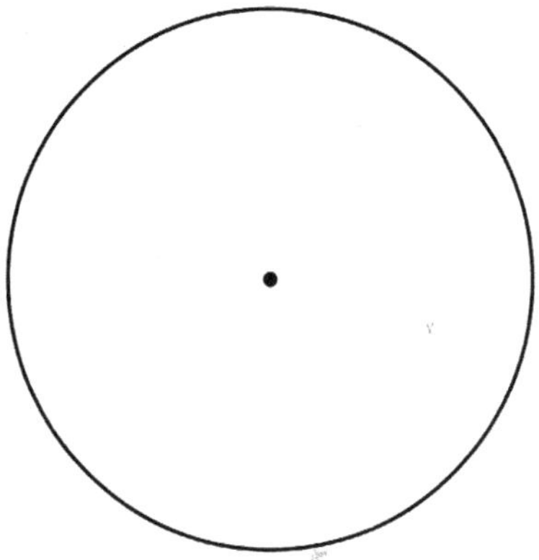

What fraction of the polygons are rhombuses? ____________

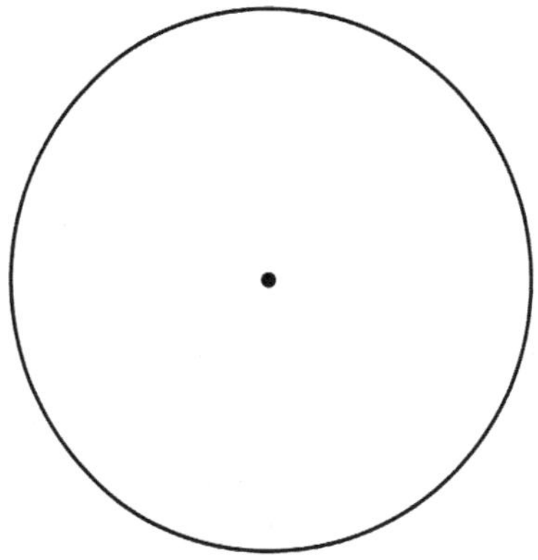

What fraction of the polygons have 4 sets of parallel lines? ____________

Name: ______________________

Date: ______________________

For the problems below, use the tangrams shown to find the percentage asked for.

Use your percentage circle to show the answer, then record it below. Shade or crosshatch the circle so that it looks like the percentage circle answer.

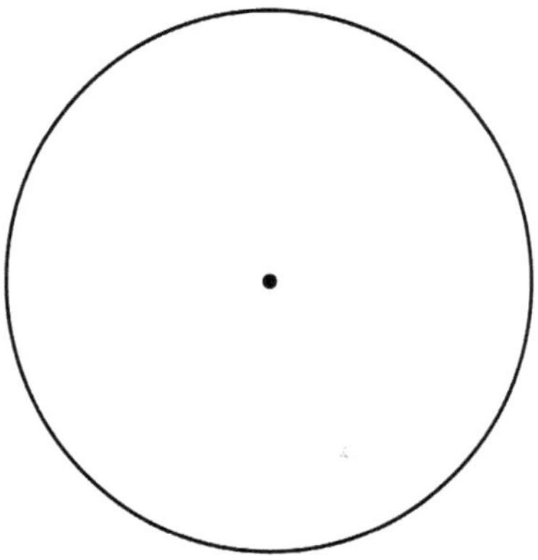

What percentage of the tangrams are in the darker color? ____________

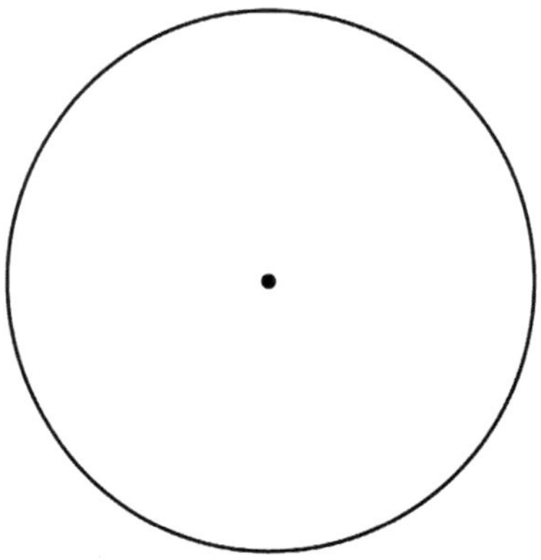

What percentage of the tangrams have right angles? ____________

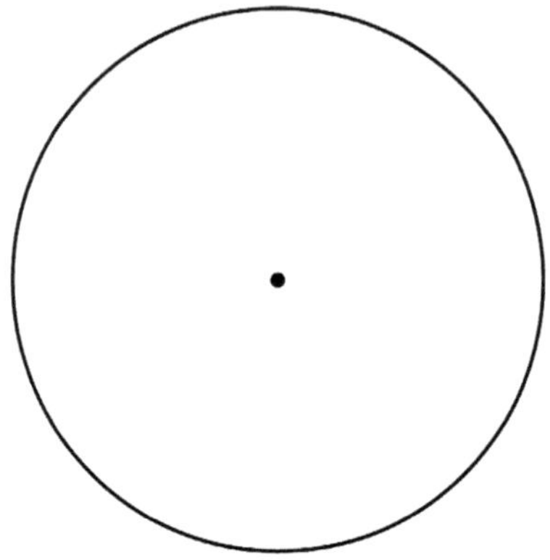

What percentage of the tangrams are triangles? ____________

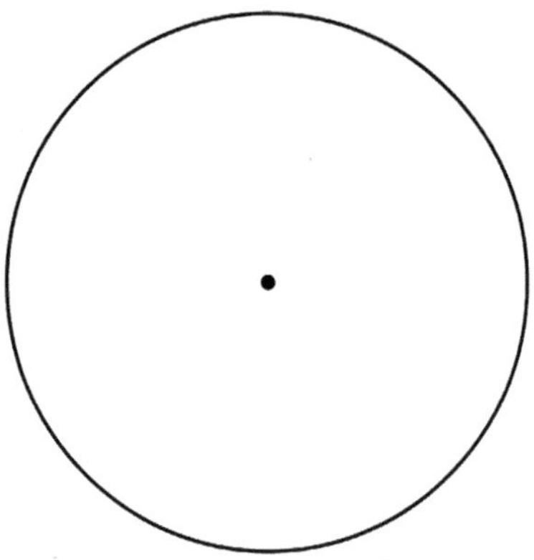

What percentage of the tangrams are quadrilaterals? ____________

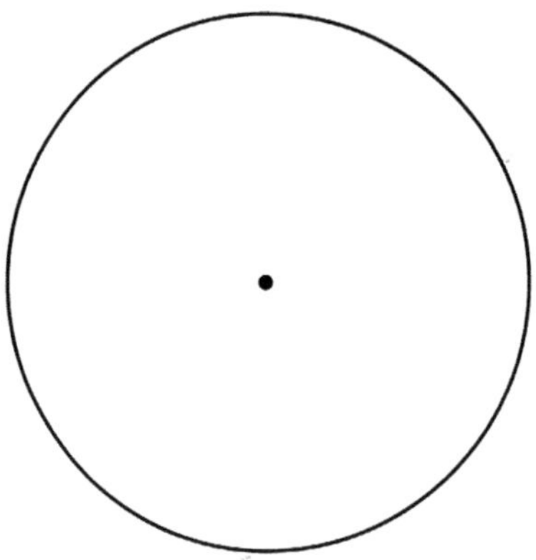

What percentage of the tangrams are polygons? ____________

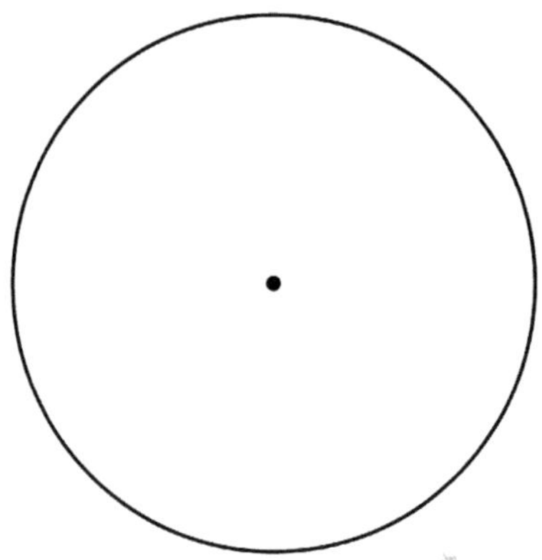

What percentage of the tangrams are rhombuses? ____________

Name: ______________________________

Date: ______________________________

For each of the five areas of the circle on the right, first guess what **fraction** of the whole circle it is. Write it near the center of the circle.

Then use your fraction circle to find the actual fraction. Write it near the edge of the circle, then add up the fractions.

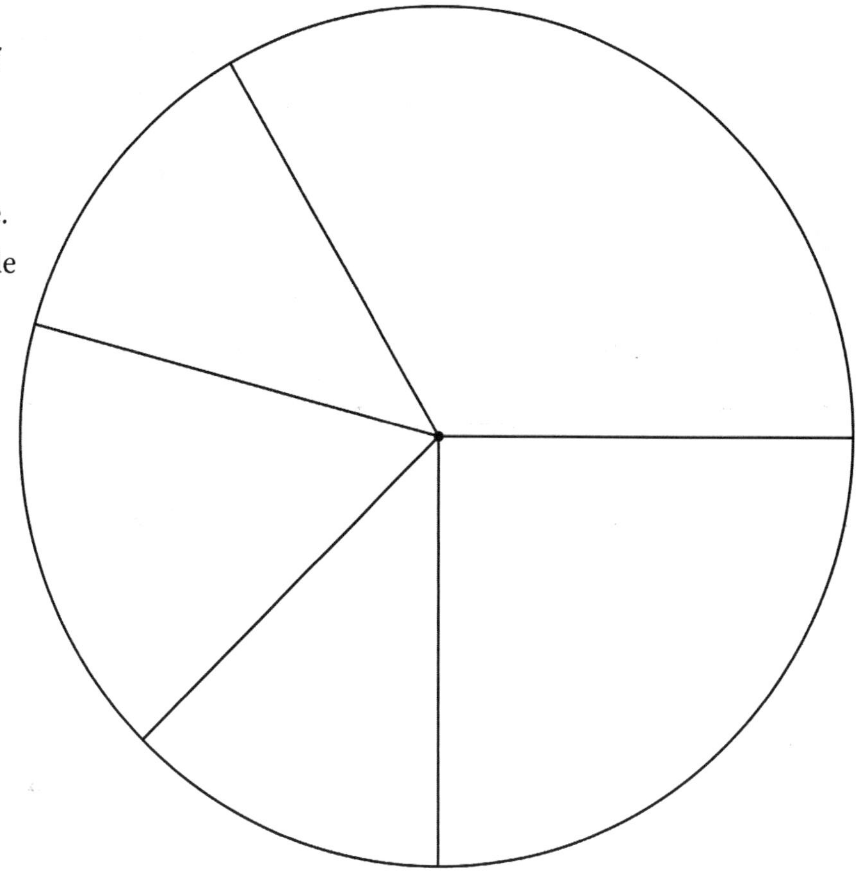

For each of the five areas of the circle on the right, first guess what **percentage** of the whole circle it is. Write it near the center of the circle.

Then use your percentage circle to find the actual percentage. Write it near the edge of the circle, then add up the percentages.

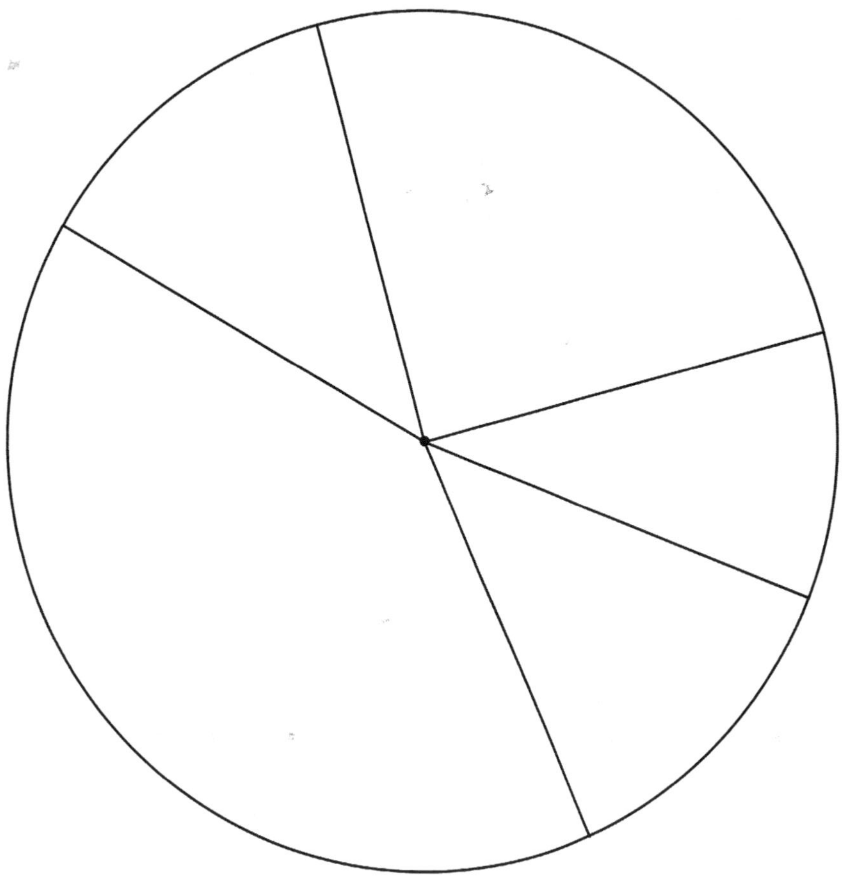

Worksheet 71–A, Review 7

Name: ______________________________

Date: ______________________________

Write only the answers.

____

____

____

Write the answers.

50% of 12 = ______

40% + ______ = 1

574 + 78 = ______

Do the calculations.

10,000 – 4682　　　　7934 + 784 + 8538　　　　2359 ÷ 7　　　　7946 × 8

Answer the questions about the tangram pieces.

What fraction of the pieces are triangles? ______

What fraction of the pieces are parallelograms? ______

What fraction of the triangles are identical? ______

What percentage of the pieces do not have a right angle? ______

What percentage of the pieces are polygons? ______

Match the fractions to the corresponding percentages.

$\frac{1}{2}$	75%
$\frac{1}{10}$	50%
$\frac{3}{4}$	100%
1	10%
$\frac{2}{10}$	20%
$\frac{1}{3}$	25%
$\frac{2}{8}$	$33\frac{1}{3}$%

Skylar had a tray of 24 tomato seedlings. Skylar gave $\frac{1}{4}$ of them to a friend, $\frac{1}{3}$ of them to another friend, and $\frac{1}{6}$ of them to a third friend. What fraction of the seedlings did Skylar have left? How many seedlings did Skylar have left?

Name: ______________________________

Date: ______________________________

Write only the answers.

____

____

____

Write the answers.

25% of 8 = _____

70% + _____ = 1

369 + 87 = _____

Do the calculations.

10,000 – 4759   6378 + 782 + 7963   7152 ÷ 8   4792 × 7

Answer the questions about the tangram pieces.

What fraction of the pieces are right triangles? _____

What fraction of the pieces are parallelograms? _____

What fraction of the pieces are rectangles? _____

What percentage of the quadrilaterals are squares? _____

What percentage of the pieces are hexagons? _____

Match the fractions to the corresponding percentages.

$\frac{1}{4}$	50%
$\frac{7}{10}$	25%
$\frac{2}{4}$	70%
1	40%
$\frac{1}{3}$	100%
$\frac{2}{5}$	75%
$\frac{3}{4}$	$33\frac{1}{3}$%

Leslie had 12 pumpkin seeds to plant. Leslie gave $\frac{1}{4}$ of them to a younger sister, $\frac{1}{6}$ of them to younger brother, and $\frac{1}{4}$ of them to a friend. What fraction of the seeds did Leslie have left to plant? How many seeds did Leslie plant?

Name: ______________________

Date: ______________________

For each figure below, calculate the angle identified by the arc. Then check it with a goniometer.

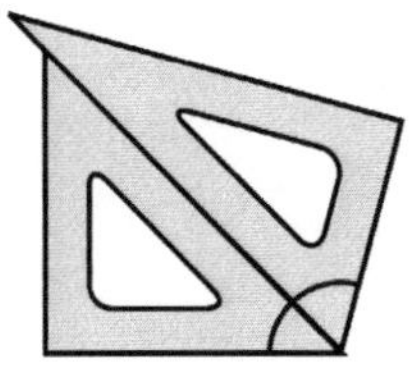

45 + 60 = 105° ✓

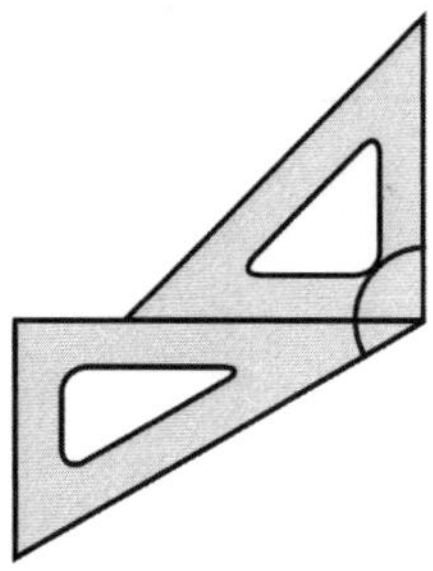

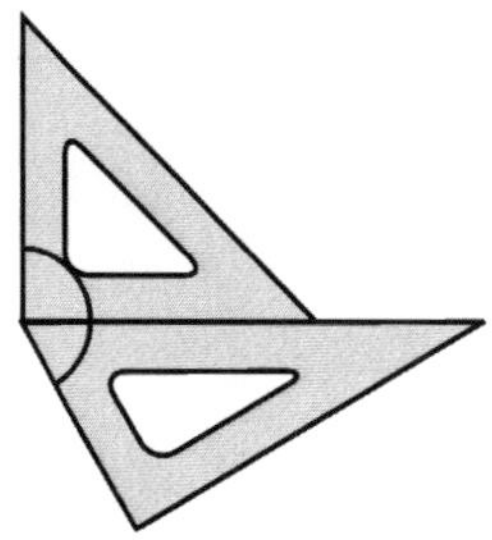

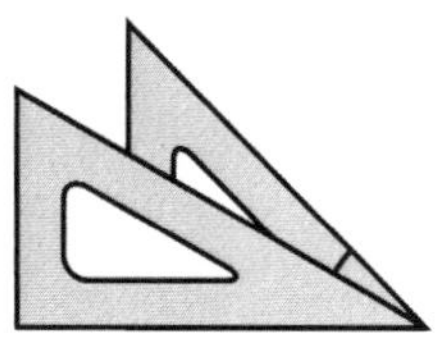

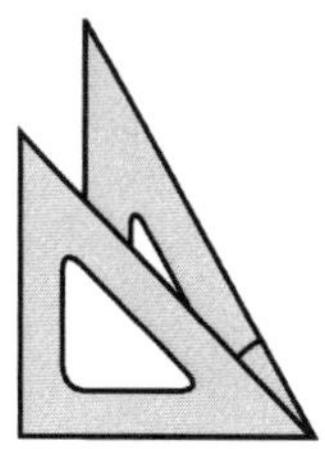

For each triangle, measure the angles and add them up.

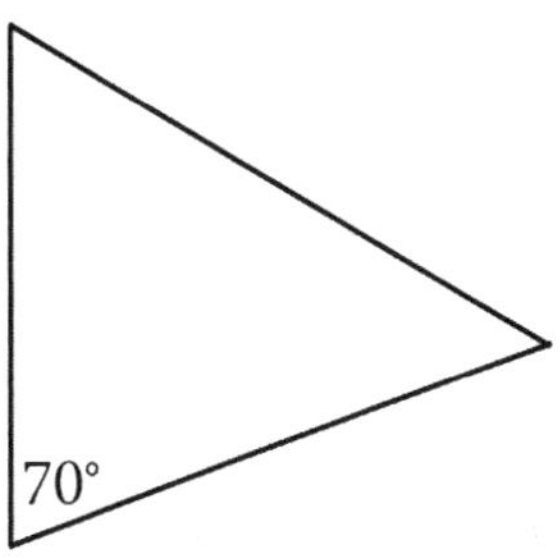

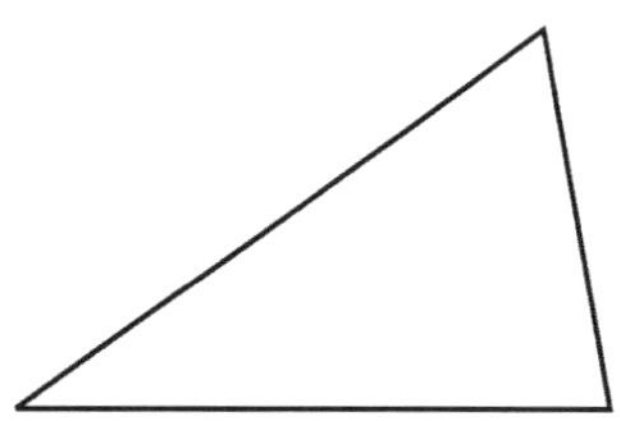

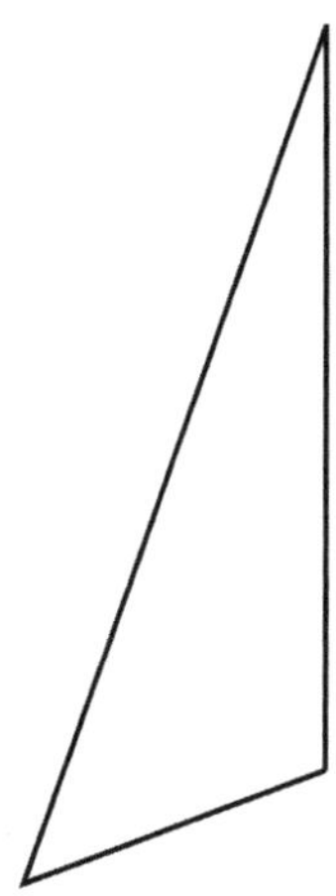

Name: ____________________ Date: ____________________

1. Measure the two base angles of the triangles below with your goniometer and the sides of the triangles in centimeters.

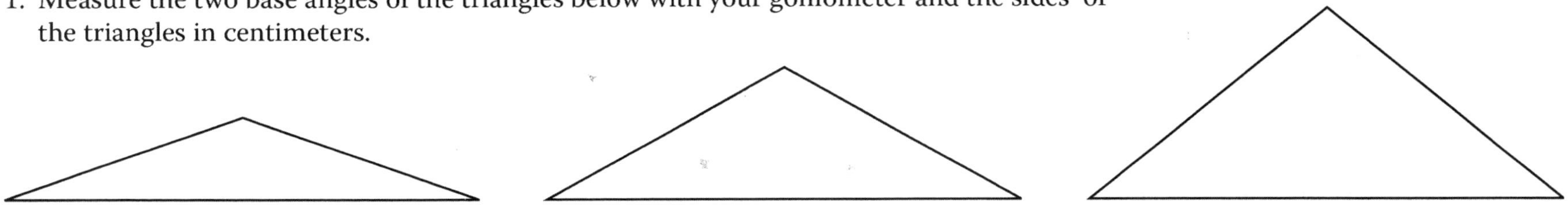

2. What do you notice about the sides opposite the congruent angles? ____________________

3. Draw a different triangle on each of these 9 × 9 peg geoboards. Each vertex of the triangle must be on a peg. Show congruent sides with little lines perpendicular to the sides. Show right angles with a little square at the 90° vertex.

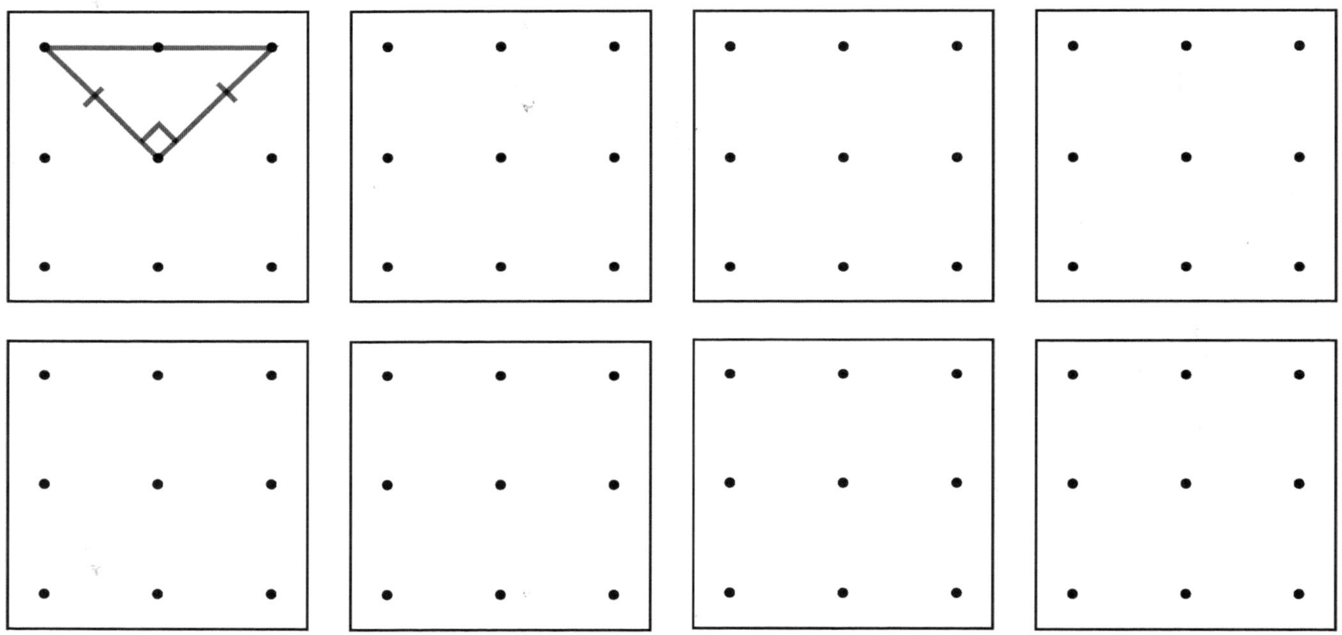

4. How many of the triangles are isosceles triangles? ________ How many of the triangles are right triangles? ________

Name: ______________________________

Date: ______________________________

Triangles can be classified by either sides or angles. Study the two tables below.

Triangles by Sides	
3 sides congruent	Equilateral
2 sides congruent	Isosceles
0 sides congruent	Scalene

Triangles by Angles	
< 90°	Acute
= 90°	Right
> 90°	Obtuse

Classify each triangle by sides and angles. Use a ruler and goniometer when needed.

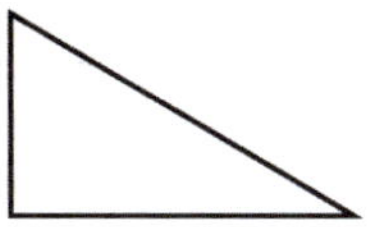
scalene ______________ ______________

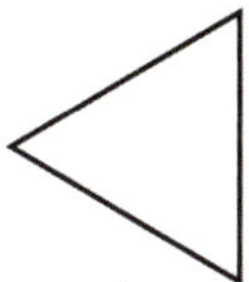
______________ ______________

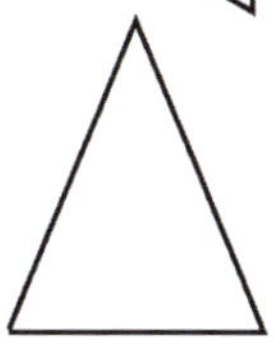
______________ ______________

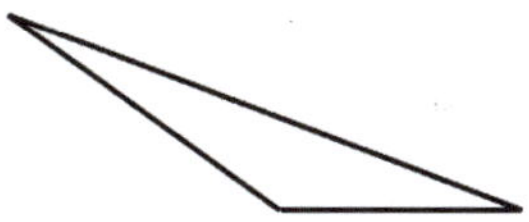
______________ ______________

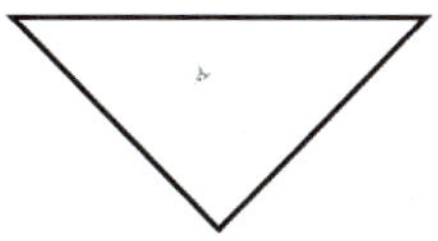
______________ ______________

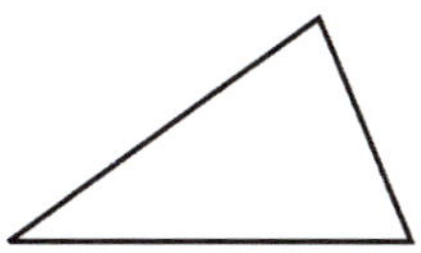
______________ ______________

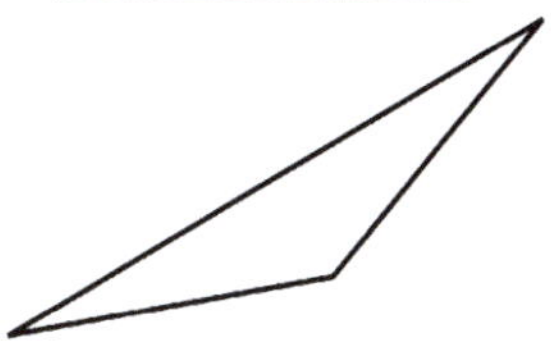
______________ ______________

Does an isosceles scalene triangle exist? Why or why not?

____

Does an equilateral right triangle exist? Why or why not?

____

Name

Date

Worksheet 75, Classifying Polygons

Name: ______________________________

Date: ______________________________

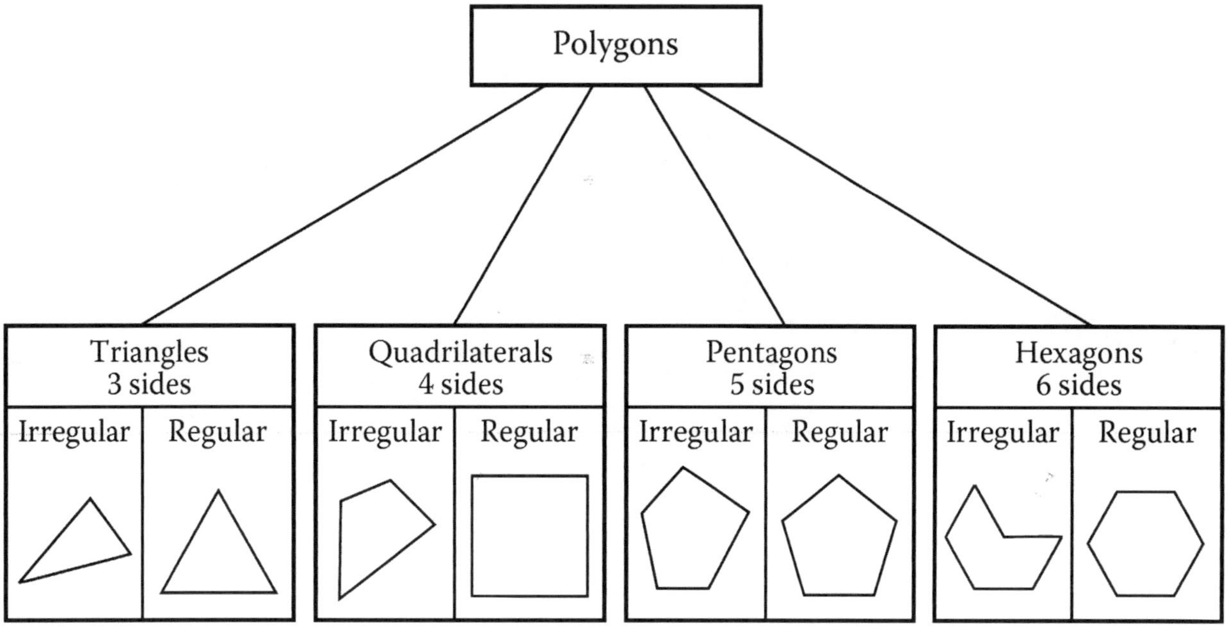

1. Which of the figures below are polygons? ______________

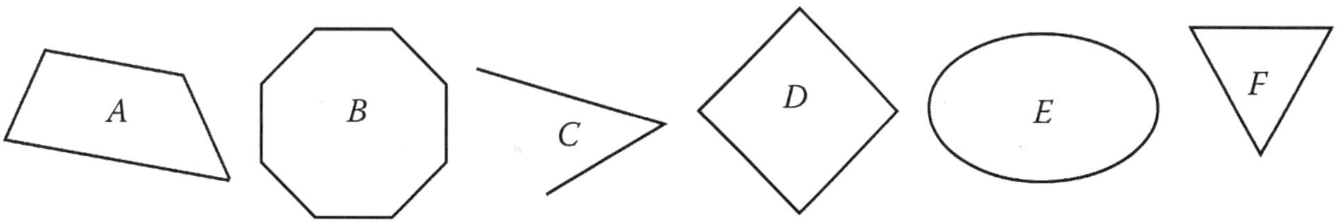

2. Explain why the others are not polygons. ______________________________

____

INFORMATION: A *regular polygon* has congruent sides and congruent angles.

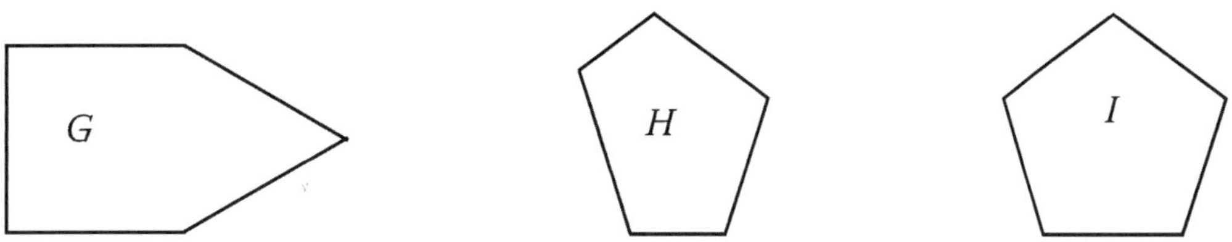

3. Which pentagons above have congruent sides? ______

4. Which pentagons above have congruent angles? ______

5. Which pentagons above are regular pentagons? ______

6. What is another name for a regular triangle? ______________________

7. What is another name for a regular quadrilateral? ______________________

8. Which of the polygons in the first question are regular polygons? ______________

Name: ______________________________

Date: ______________________________

3. Use drawing tools and a goniometer to construct a protractor like the one shown. Label only the multiples of 10. Then cut out the protractor. Also cut out the shaded area inside.

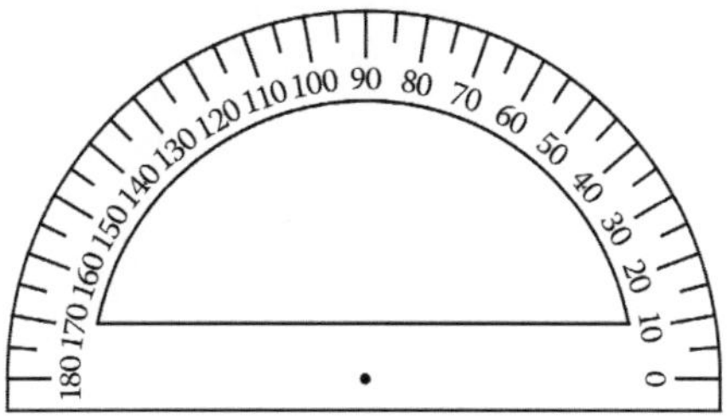

A protractor.
The point on the protractor is the vertex of the angle.

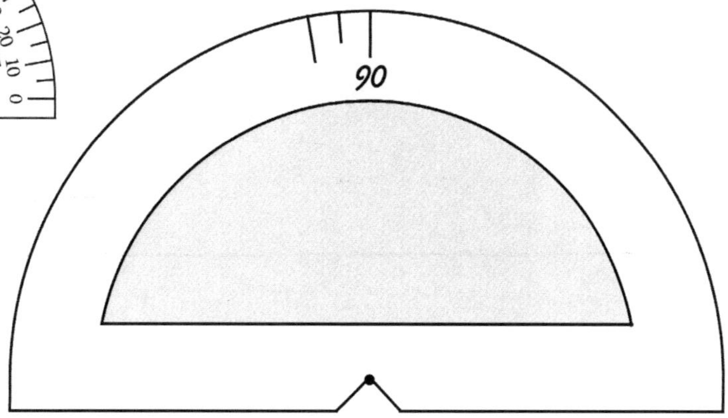

2. Cut out the circle and cut on the line that goes to the center.

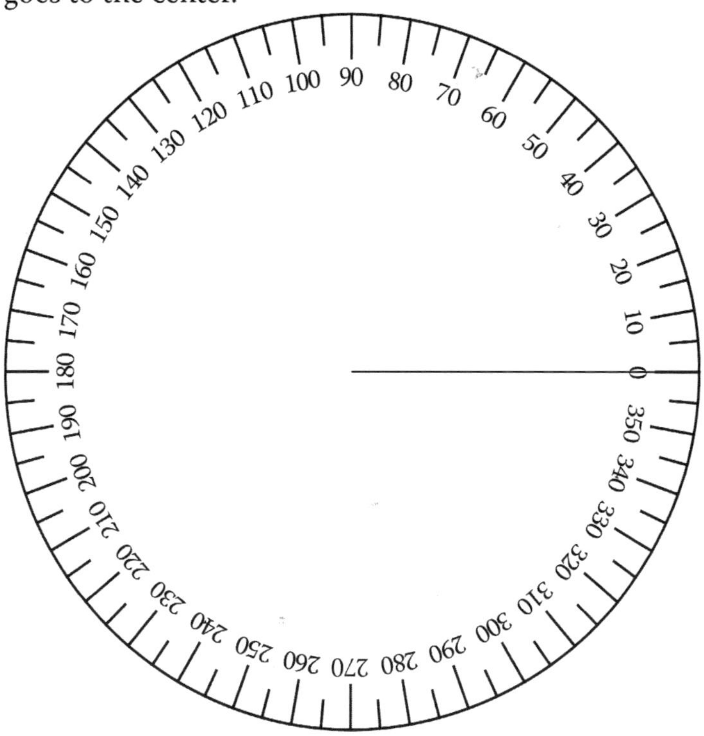

1. Cut out the triangles and save them for the next lesson.

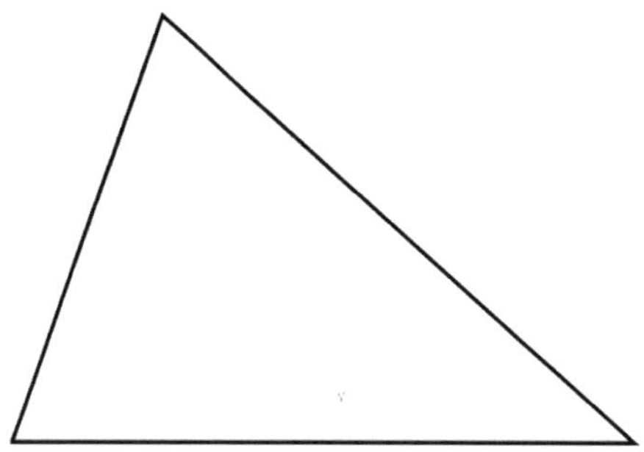

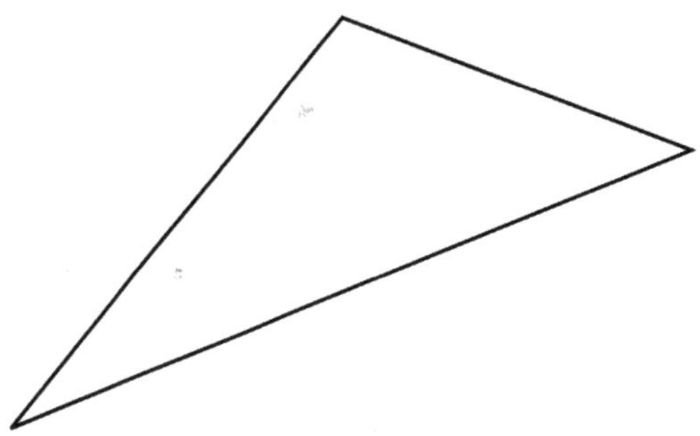

Worksheet 77-1, Angles in a Circle

Name: ____________________

Date: ____________________

For each circle, measure the angles with the protractor from the last lesson. Then add the angles in a circle.

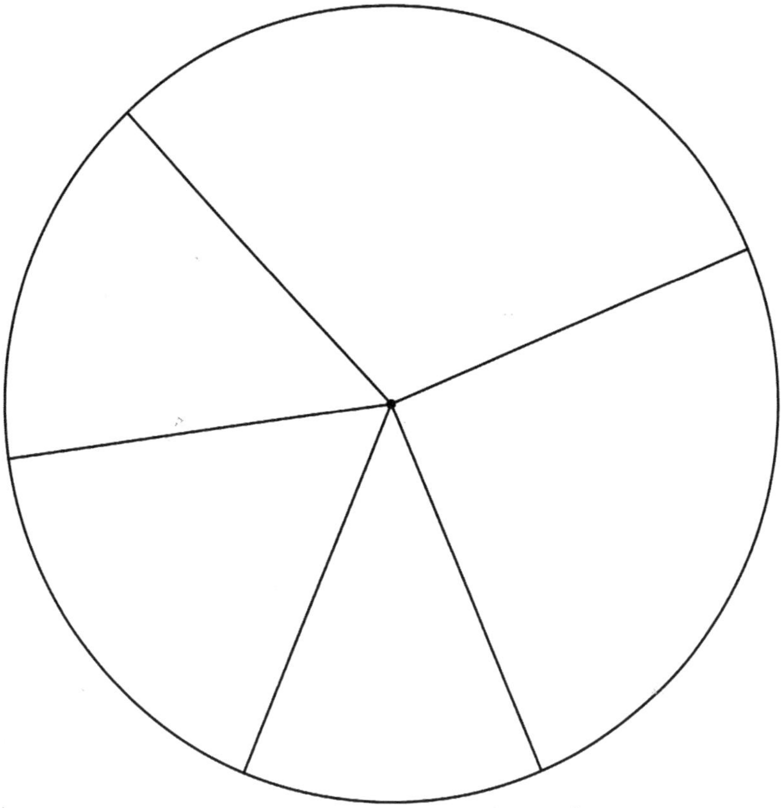

Sum of angles in the circle is ________

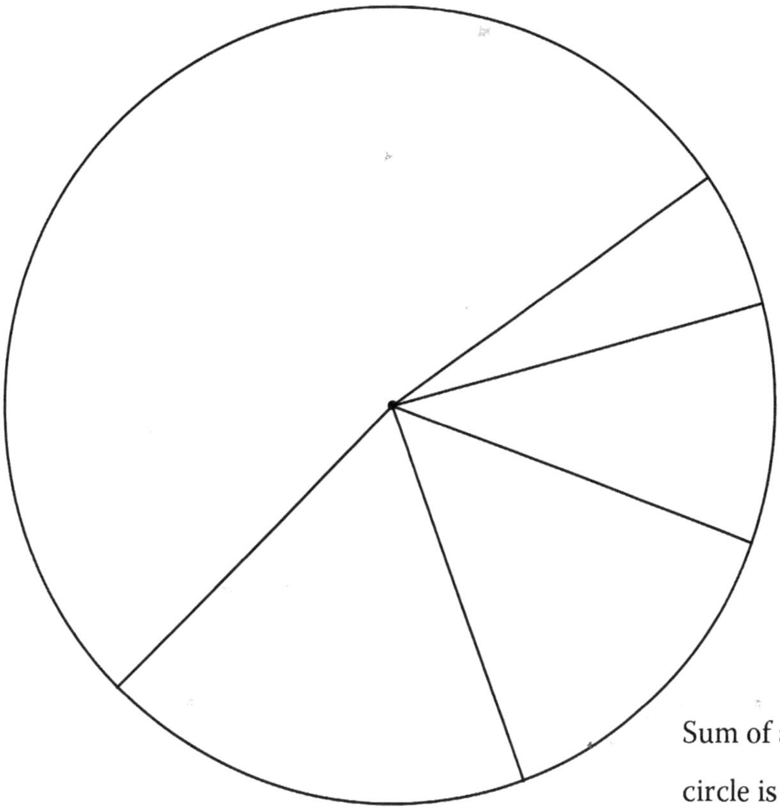

Sum of angles in the circle is ________

Name: ______________________________

Date: ______________________________

For each clock, draw the angle and find the angle the hand travels in degrees.

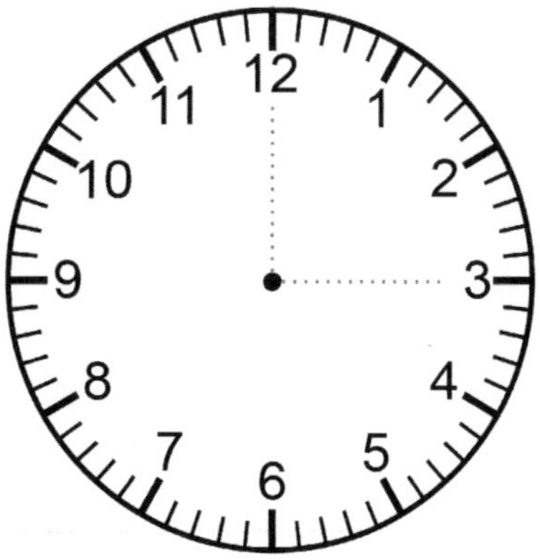

From 12 to 3: ________

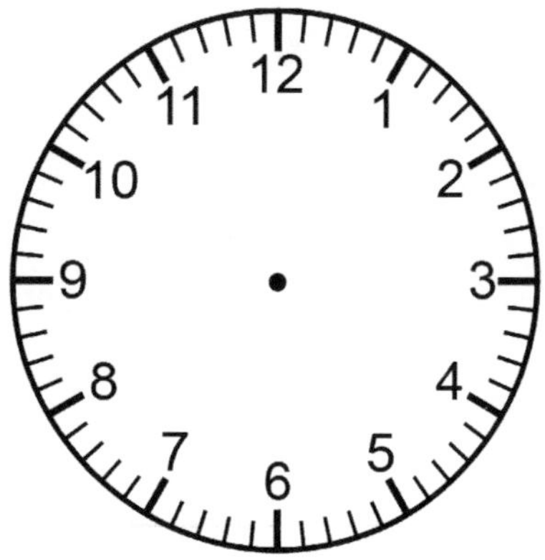

From 12 to 1: ________

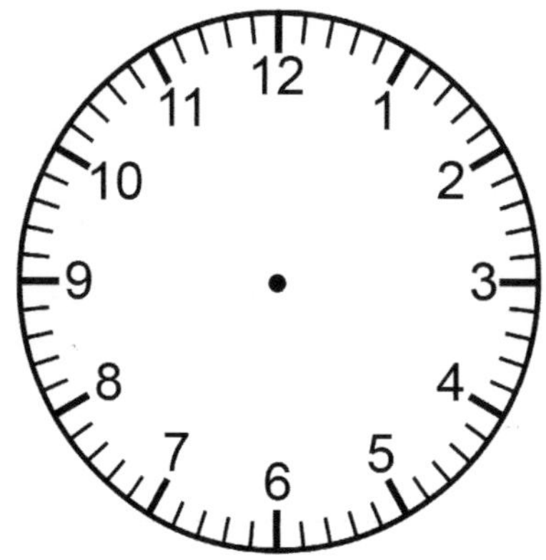

From 12 to 2: ________

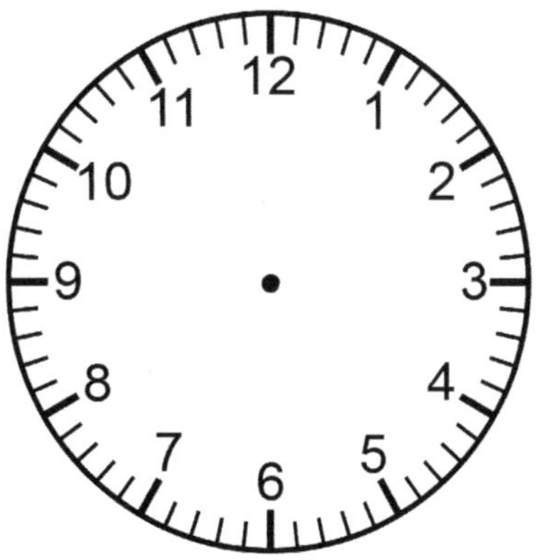

From 12 to 6: ________

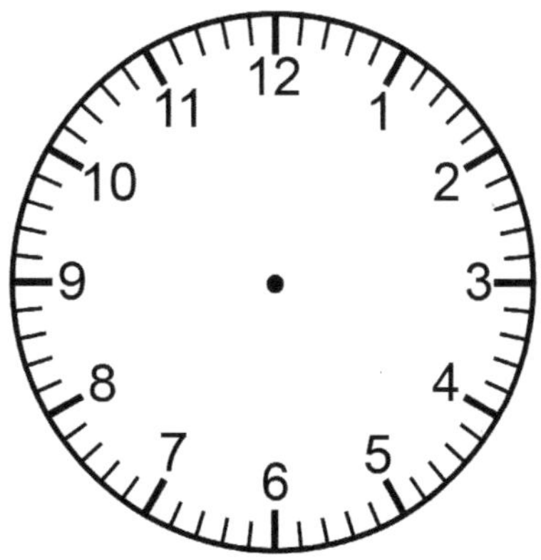

From 3 to 12: ________

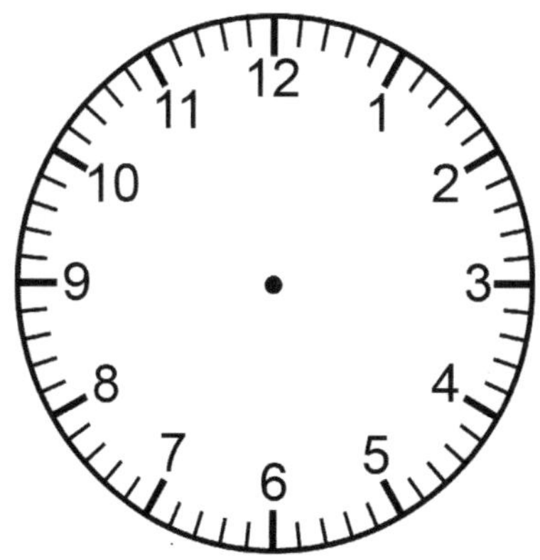

From 1 to 8: ________

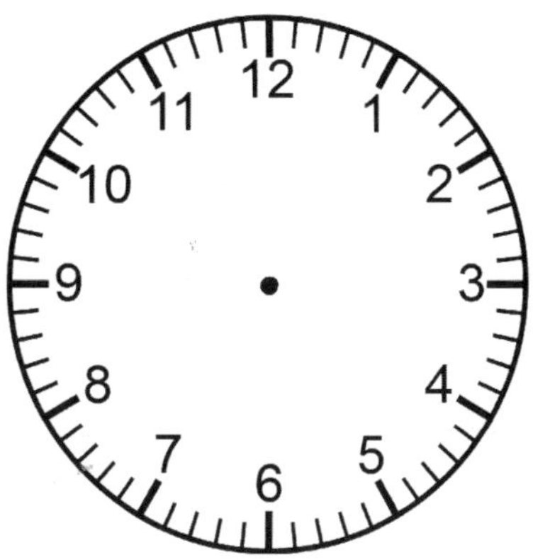

From 6 to 10: ________

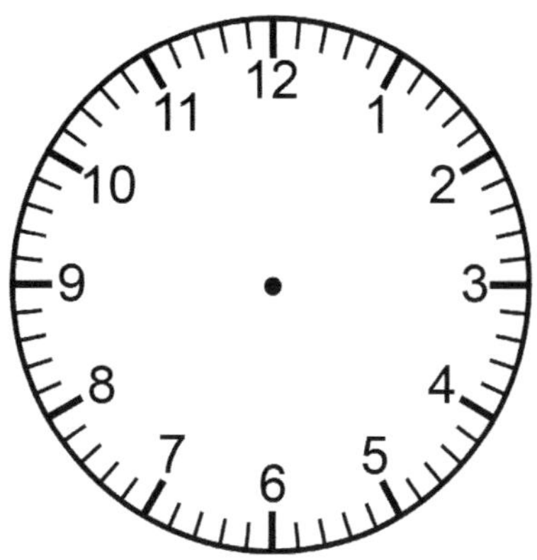

From 7 to 12: ________

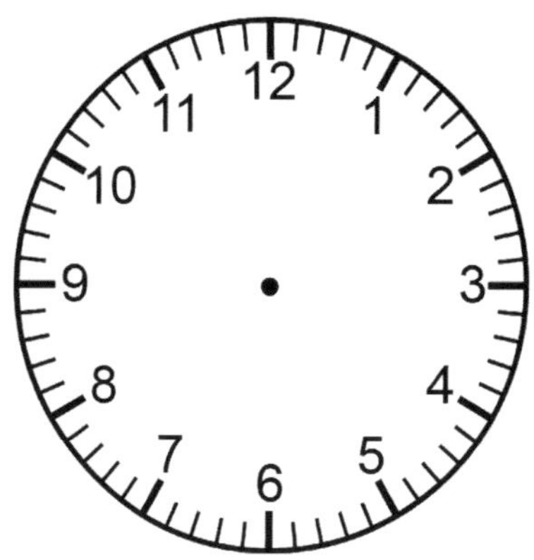

From 7 to 3: ________

Worksheet 78, Angles on a Geoboard

Name: ____________________

Date: ____________________

INFORMATION: An *inscribed angle* has its vertex on the circle.
A *central angle* has its vertex at the center of a circle.

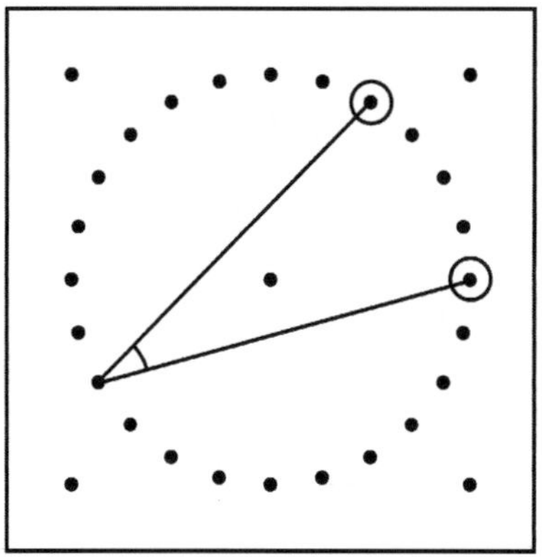

Measure the inscribed angle. ________

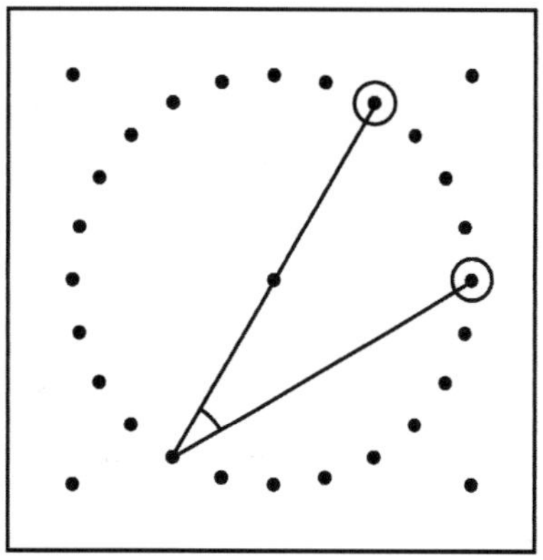

Measure the inscribed angle. ________

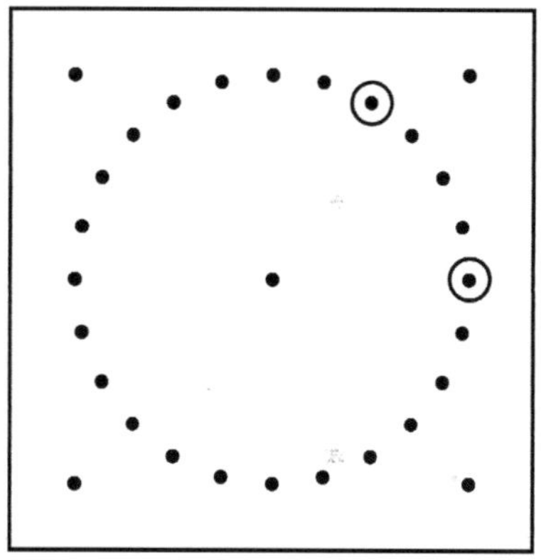

Move the vertex to another peg. Measure it. ________

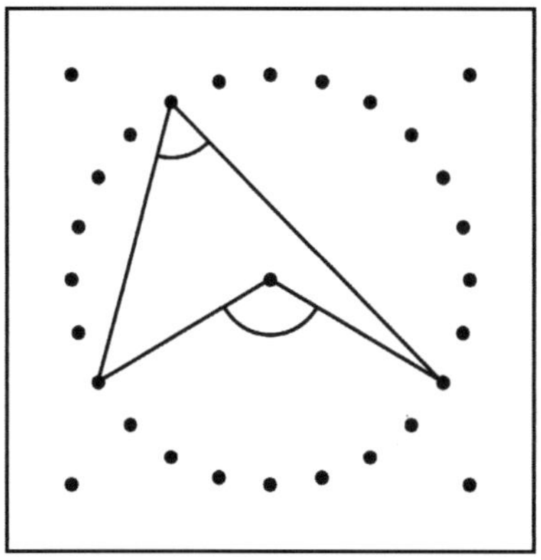

Measure the two angles. ____________

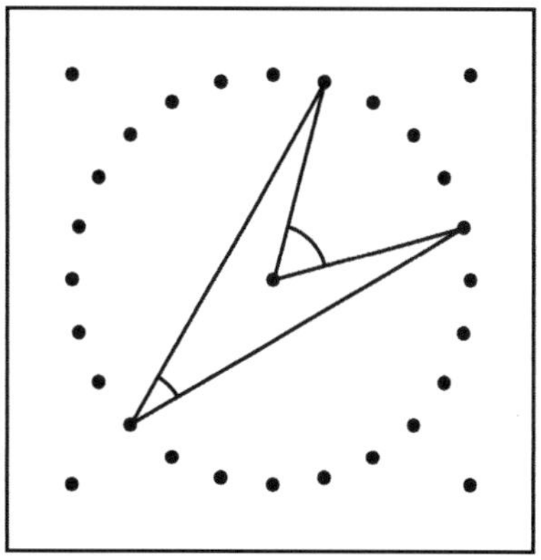

Measure the two angles. ____________

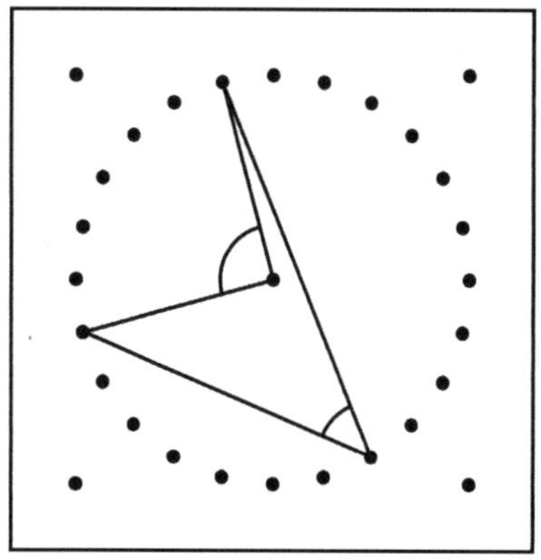

Measure the two angles. ____________

What is the relationship between the inscribed angle and the central angle? __________

____

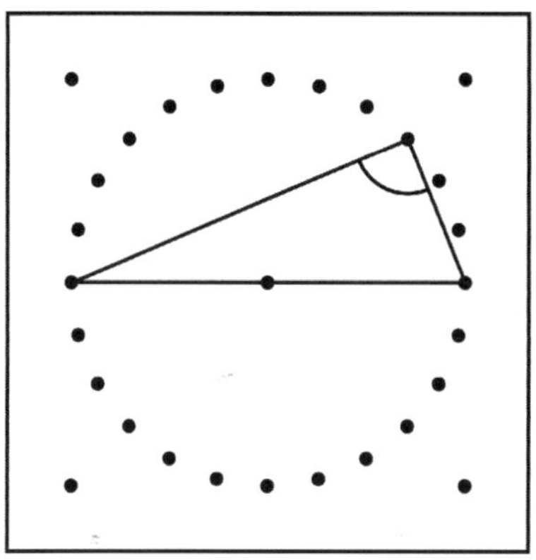

Measure the inscribed angle. ________

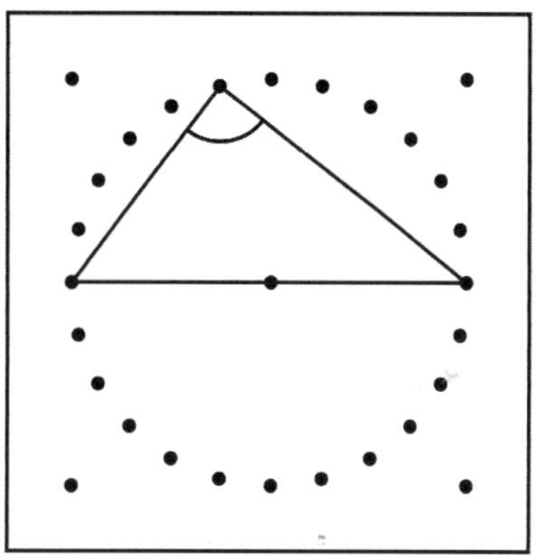

Measure the inscribed angle. ________

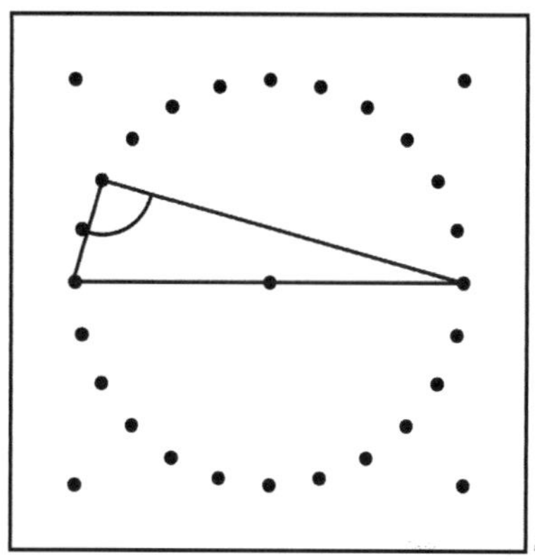

Measure the inscribed angle. ________

Worksheet 79, Regular Polygons on a Geoboard

Name: ______________________________

Date: ______________________________

Use a ruler or straightedge to draw the figures below.

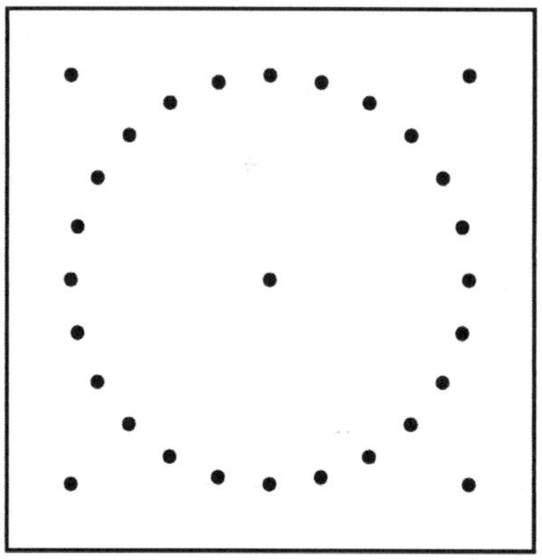

Draw the central angle for a hexagon.

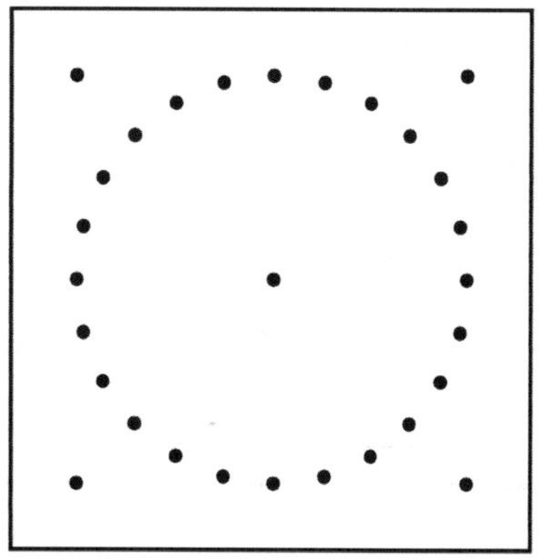

Draw a regular hexagon.

The central angle is ________

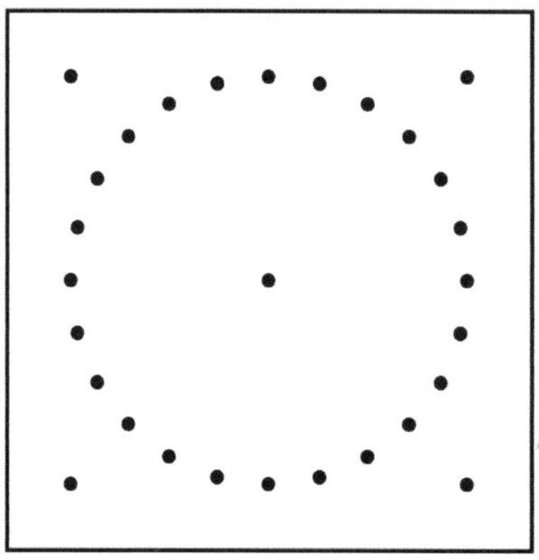

Draw a square. The central angle is ________

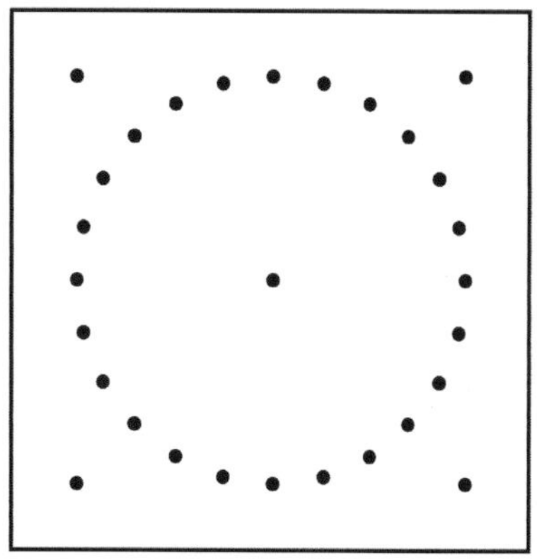

Draw an equilateral triangle.

The central angle is ________

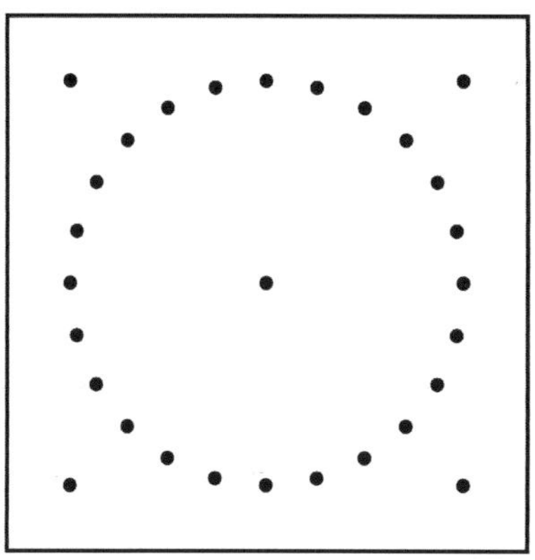

Draw a regular octagon. The central angle is ________

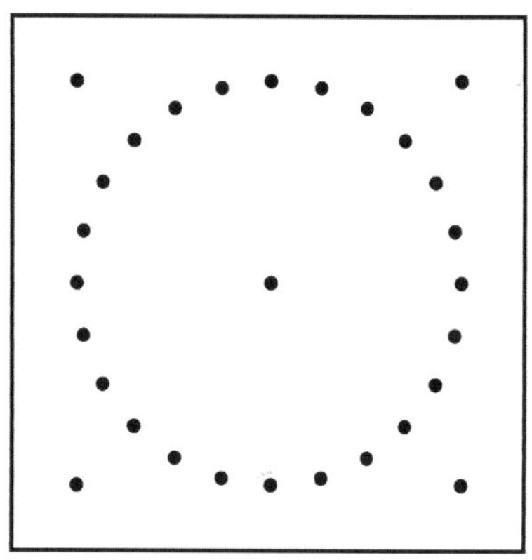

Draw a regular dodecagon, a 12 sided polygon.

The central angle is ________

Worksheet 80-A, Review 8

Name: ____________________

Date: ____________________

Write only the answers.

____

____

____

Write the answers.

50% of 16 = ______

55% + ______ = 1

693 + 58 = ______

Do the calculations.

12,345 – 9483 | 29 + 784 + 8538 | 5932 ÷ 7 | 6582 × 9

Find the missing angle.
Write an equation.

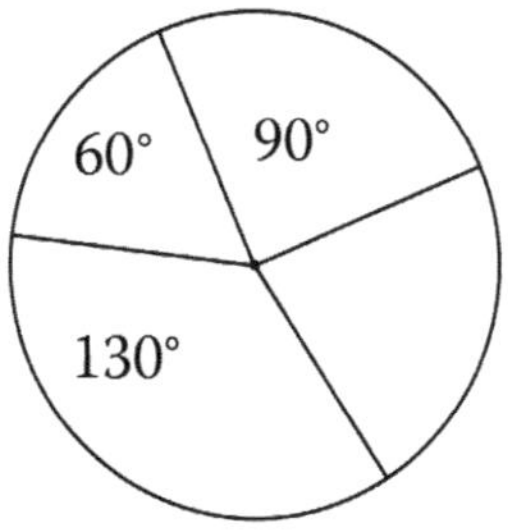

____

____

Write T for true or F for false before each statement.

____ An equilateral triangle has equal angles.

____ A rhombus has equal sides.

____ A square is not a regular polygon.

____ The central angles of a regular polygon are congruent.

____ A hexagon with congruent sides is always a regular polygon.

____ A circle is a regular polygon.

____ An irregular polygon cannot have congruent angles.

Classify the triangles. Classifications can have more than one answer.

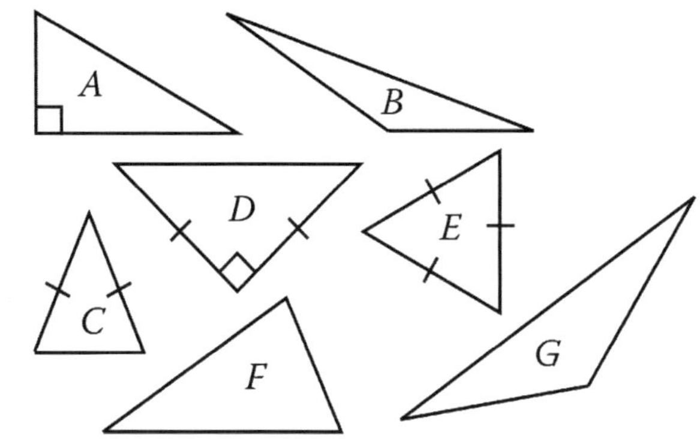

Equilateral ____________

Isosceles ____________

Scalene ____________

Acute ____________

Right ____________

Obtuse ____________

Which angle is greatest?

____

Name: ____________________

Date: ____________________

Write only the answers.

____

____

____

Write the answers.

50% of 18 = ______

35% + ______ = 1

768 + 78 = ______

Do the calculations.

13,574 – 5968   17 + 904 + 6587   2759 ÷ 8   7586 × 7

Find the missing angle.
Write an equation.

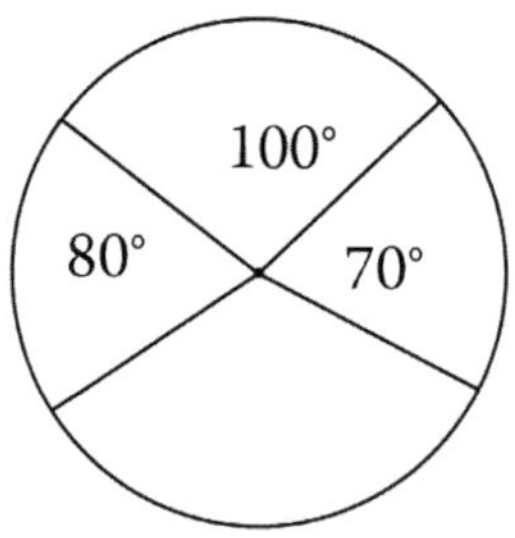

____

____

Write T for true or F for false before each statement.

____ An equilateral triangle has equal sides.

____ A rhombus always has equal angles.

____ A rectangle can never be a regular polygon.

____ The least number of sides a polygon can have is three.

____ A hexagon with congruent angles is always a regular polygon.

____ A circle is a special polygon.

____ A trapezoid is an irregular polygon.

Classify the triangles. Classifications can have more than one answer.

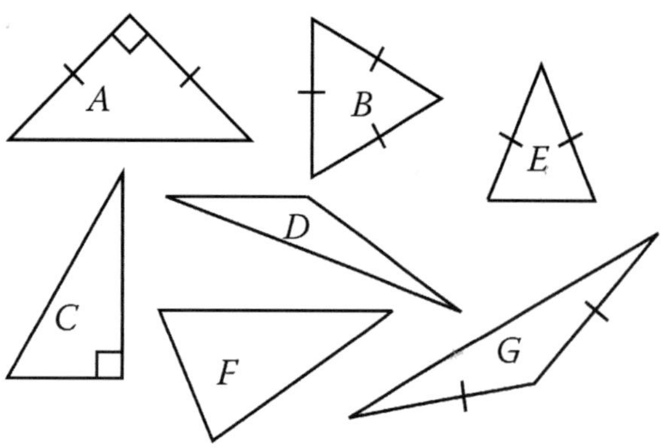

Acute ____________________

Right ____________________

Obtuse ____________________

Equilateral ____________________

Isosceles ____________________

Scalene ____________________

Which angle is least?

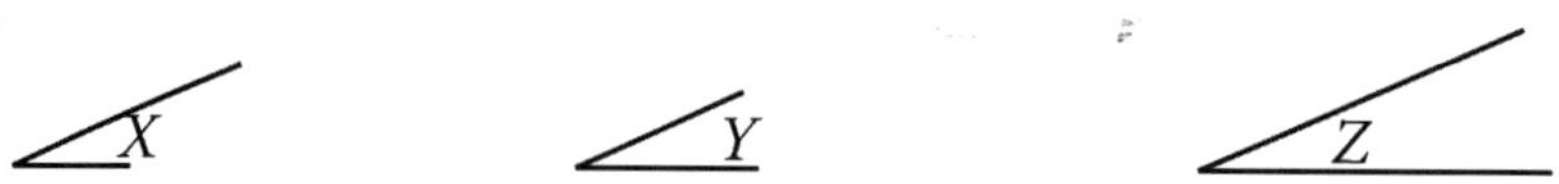

____

Name: ______________________

Date: ______________________

Warm-Up

Multiply.

$$\begin{array}{r} 5\ 2\ 3 \\ \times\ 6\ 8 \\ \hline \end{array} \quad (\ )\ (\ )\ (\ )$$

$$\begin{array}{r} 2\ 3\ 4\ 6 \\ \times\ 9\ 3 \\ \hline \end{array} \quad (\ )\ (\ )\ (\ )$$

Do the calculations below.

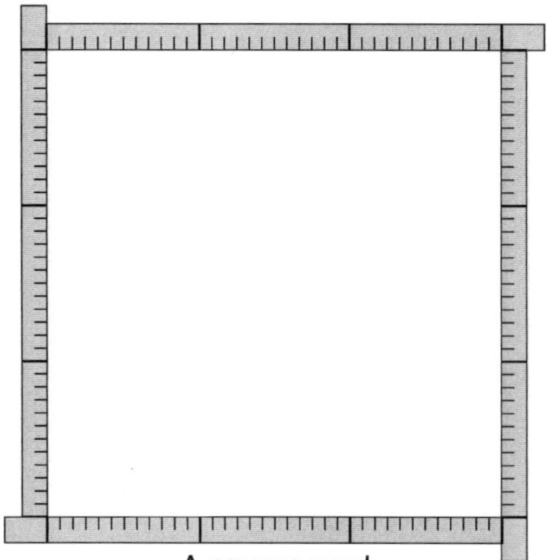

A square yard

A square meter

Draw a square foot (sq ft) in the figure.

How many feet are in a yard? ________

How many sq ft in a sq yard? ________

Draw a sq in. in the sq ft.

How many inches in a foot? ________

How many sq in. in a sq ft? ________

How many inches in a yard? ________

How many sq in. in a sq yard? ________

Draw a square decimeter (dm^2) in the figure.

How many decimeters in a meter? ________

How many dm^2 in a square meter (m^2)? ________

Draw a square centimeter (cm^2) in the dm^2.

How many centimeters in a decimeter? ________

How many cm^2 in a dm^2? ________

How many centimeters in a meter? ________

How many cm^2 in a m^2? ________

Name: ______________________________

Warm-Up

Date: ______________________________

Multiply.

```
   6 1 8 ( )          3 6 1 8 ( )
 ×   4 9 ( )            ×   5 2 ( )
 ___________        _____________
         ( )                  ( )
```

Solve these problems. Make a drawing for Problems 2 to 4.

1. Which desk is larger, one that is 13 dm long by 11 dm wide, or one that is a square measuring 1 m and 2 dm on a side?

2. A room is 117 m^2. One side is 9 m long. How long is the other side?

3. A room is 18 ft by 18 ft. What will the carpet cost if carpeting is $11.95 per square yard?

4. Jacob measured a bedroom and found it was 9 ft by 15 ft. What will the carpet cost if at $9.95 per square yard?

5. Isabel measured a hallway. It measures 36 ft by 36 inches. Hallway carpeting costs $8.50 a square yard. The installer charges $3.25 a square yard. What is the total cost?

Name: ______________________________

Date: ______________________________

Warm-Up

Multiply.

```
   2 4 9 ( )          6 2 4 9 ( )
 ×   2 7 ( )        ×       3 8 ( )
 ———————————        ———————————————
           ( )                    ( )
```

Fill in the blanks.

Miles (mi)	Yards (yd)	Feet (ft)
1	1760	5280
2		
7		
		528
$\frac{1}{2}$		
$\frac{1}{4}$		
n	1760 × n	

Kilometers (km)	Meters (m)
1	1000
2	
3	
4	
10	
	500
n	

Solve the problems. First, convert all the units into the smaller units. Then convert the sum into the larger unit.

1. McKenzie ran 880 yards on Monday, 2 miles on Tuesday, 440 yards four times on Wednesday, and 1.5 miles on Thursday. How many yards did she run during the week? How many miles did she run?

2. Sydney raced on her unicycle 400 m around a track on Monday, twice around the same track on Tuesday and Wednesday, and rode 2 km on Thursday. How many meters did she ride during the week? How many kilometers did she ride?

Name: ____________________

Date: ____________________

Gallon, half gallon, and quart containers.

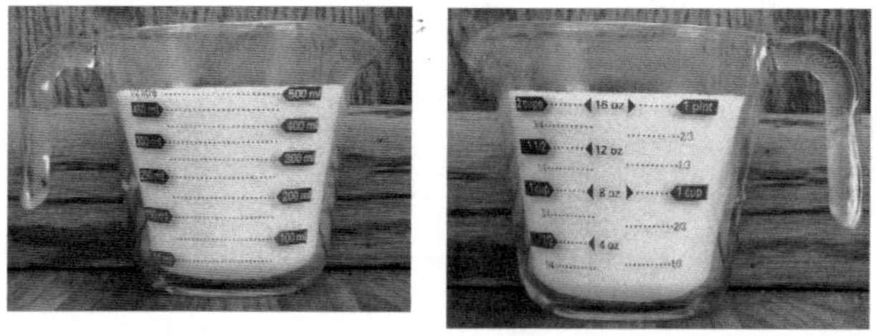

Measuring cup showing cups and ounces on one side and milliliters on the other.

Fill in the blanks.

Gallons	Quarts (qt)	Cups
1	4	
	1	4
1		
2		
$\frac{1}{2}$		
		12
	16	
n		

Pints	Cups	Ounces (oz)
1	2	
	1	8
1		
2		
		24
	6	
		64
n		

Liters (L)	Milliliters (ml)
1	1000
2	
	3000
0.5	
1.5	
	100
	200
n	

Solve these problems.

1. Eileen is making a punch with a 12-oz can of lemonade, 4 cups cranberry juice, and 1 qt of soda. How many cups of punch will this recipe make? How many servings will there be if each serving is 4 oz?

2. Zane is mixing a drink using 2 cans with 275 ml frozen cranberry concentrate, 1 L of orange juice, 60 ml of lime juice, and 1.5 L of soda. Will there be enough of the drink for 17 people to each have 180 ml?

Name: ____________________

Date: ____________________

Warm-Up

Multiply.

```
   2 7 5 ( )          7 2 7 5 ( )
 × 6 3 ( )              × 7 4 ( )
       ( )                    ( )
```

Fill in the blanks.

Pound (lb)	Ounces (oz)
1	16
2	
7	
75	
76	
$\frac{1}{4}$	
$\frac{1}{2}$	
n	

Ton	Pound (lb)
1	2000
2	
5	
10	
n	

Kilogram (kg)	Gram (g)
1	1000
2	
3	
	6000
	4000
	100
	500
n	

Solve these problems.

1. Ryan's bicycle weighs 15 pounds. His car weighs 200 times that much. How many pounds does his car weigh? How many tons does his car weigh?

2. Each little centimeter cube weighs 1 gram. One hundred cubes come in a package. How many packages do you need to weigh 10 kilograms?

Name: ____________________

Date: ____________________

Warm-Up

Multiply.

```
  4 5 1 ( )          8 4 5 1 ( )
 ×  8 6 ( )            × 9 7 ( )

        ( )                  ( )
```

Fill in the blanks.

Day	Hour (hr)	Minute (min)
1	24	
	1	60
1		
2		
	12	
7		
365		
n		

Minute (min)	Second (sec)	Millisecond (msec)
1	60	
	1	1000
1		
		500
2		
		6000
60		
n		

Solve these problems.

1. Kenna spends 5 hours a day in school Monday to Friday. How many hours is that in a week? How many hours during a week is she not in school?

2. Dez had a job for a year (52 weeks) except for 2 weeks of vacation. She worked 8 hours a day 5 days a weeks. How many hours did she work in a year?

Name: ______________________________

Date: ______________________________

Make line plots for two different years, one not a leap year and one a leap year.

First Day of the Months in a Non-leap Year

S M T W Th F S

First Day of the Months in a Leap Year

S M T W Th F S

Which is the most common day for the first of the month? ______________

Which is the most common day for the first of the month? ______________

Make a line plot using the dimensions in the tables below. Show the fraction divisions on the line.

Geometric Solid	Edge or Diameter
cone	$1\frac{7}{8}$
cube	2
cylinder, short	$1\frac{7}{8}$
cylinder, long	1
hemisphere	$1\frac{7}{8}$
hexagonal prism	3

Geometric Solid	Edge or Diameter
octagonal prism	3
pyramid	$3\frac{1}{4}$
rectangular prism	$1\frac{3}{4}$
square prism	3
triangular prism	3

1 2 3

What is the difference between the most frequent dimension and the next most frequent dimension?

____

Name: ______________________

Date: ______________________

Write only the answers.

____

____

____

1. After multiplying by 16, 24, 27, and 35, divide by 2, 3, 4, 5, 6, 7, 8, and 9.

```
   1 2  ( )
 × 1 6  ( )
 _____

        ( )
 × 2 4  ( )
 _____

        ( )
 × 2 7  ( )
 _____

        ( )
 × 3 5  ( )
 _____
        ( )
        ( )
        ( )
        ( )
        ( )
        ( )
        ( )
        ( )
```

Write the answers.

How many oz in half a lb? (1 lb = 16 oz) = _____

$4 + $1.25 = ______

(50 × 6) – (49 × 6) = ______

2. Below are the heights of the 2016 Minnesota Lynx Women's Basketball team. What is the difference in height between the shortest and the tallest players?

____

Player	Height
Seimone	6 ft 0 in.
Rebekkah	6 ft 2 in.
Sylvia	6 ft 6 in.
Bashaara	6 ft 2 in.
Keisha	6 ft 1 in.
Natasha	6 ft 3 in.

Player	Height
Janel	6 ft 2 in.
Renee	5 ft 7 in.
Maya	6 ft 0 in.
Jia	5 ft 8 in.
Lindsay	5 ft 9 in.

3. Logan is serving cranberry juice to senior citizens. He has a gallon and a quart of juice. Each serving is half a cup. How many servings can he make? (1 quart = 4 cups)

4. Paxton wants to buy enough orange juice for each team member to have 250 milliliters to drink. There are 18 team members. How many 2-L containers does he need to buy? How much will be left over?

Name: ______________________________

Date: ______________________________

Write only the answers.

____

____

____

Write the answers.

How many oz in $\frac{1}{4}$ of a lb? (1 lb = 16 oz) = _____

\$3.65 + \$5 = _____

(40 × 5) – (39 × 5) = _____

1. After multiplying by 16, 24, 27, and 35, divide by 2, 3, 4, 5, 6, 7, 8, and 9.

2 1	( )
× 1 6	( )
	( )
× 2 4	( )
	( )
× 2 7	( )
	( )
× 3 5	( )
	( )
	( )
	( )
	( )
	( )
	( )
	( )
	( )

2. Below are the heights of the 2016 New York Liberty Women's Basketball team. What is the difference in height between the shortest and the tallest players?

____

Player	Height
Brittany	5 ft 9 in.
Adut	6 ft 4 in.
Swin	6 ft 1 in.
Tina	6 ft 4 in.
Epiphanny	5 ft 9 in.
Sugar	5 ft 9 in.

Player	Height
Shoni	5 ft 9 in.
Kiah	6 ft 3 in.
Carolyn	6 ft 6 in.
Tanisha	5 ft 11 in.
Amanda	6 ft 5 in.
Shavonte	5 ft 10 in.

3. Sebastian is serving pomegranate juice to senior citizens. He has a gallon and a half of juice. Each serving is half a cup. How many servings can he make? (1 quart = 4 cups)

4. James is buying enough orange juice for each team member to have 250 milliliters to drink. There are 21 team members. How many 2-L containers does he need to buy? How much will be left over?

Worksheet 89, Shapes in an Octagon

Name: ______________________________

Date: ______________________________

First make the figures on your geoboard. Then draw the figures below either freehand, or with a ruler or straightedge. All lines must connect vertices on the octagon.

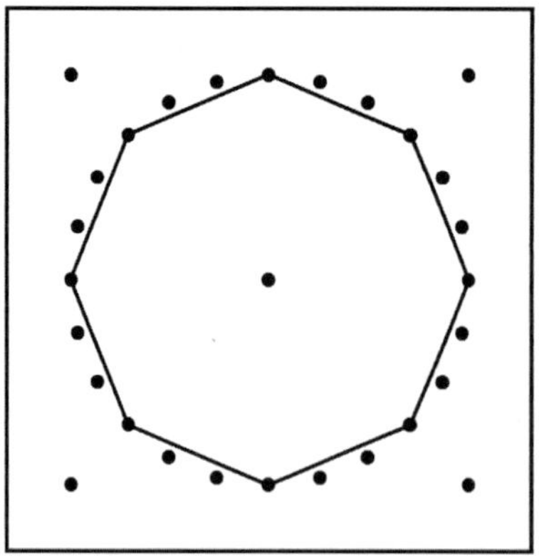

Draw a square.

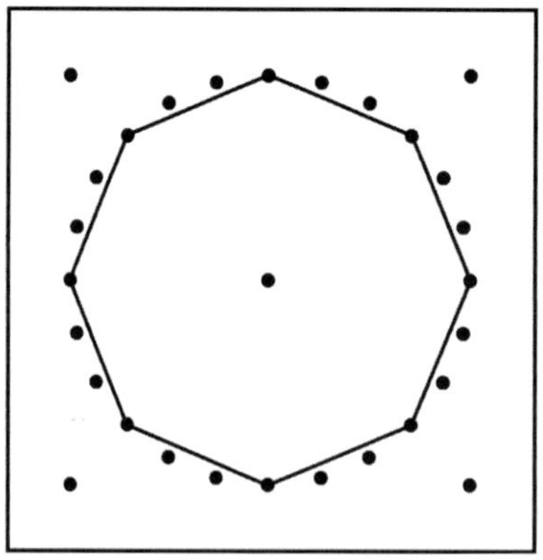

Draw a rectangle that is not a square.

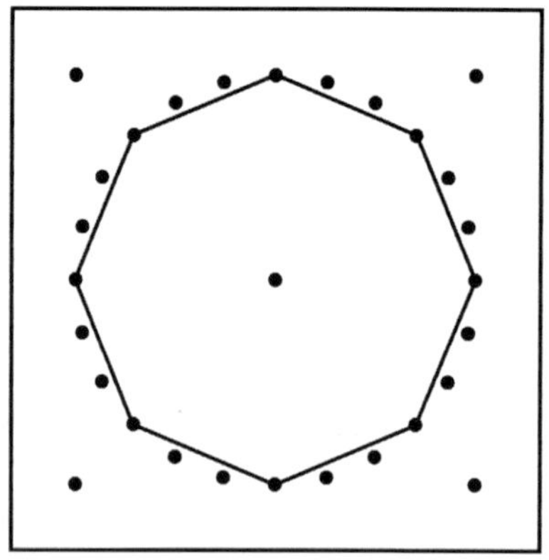

Draw two lines to make a right triangle.

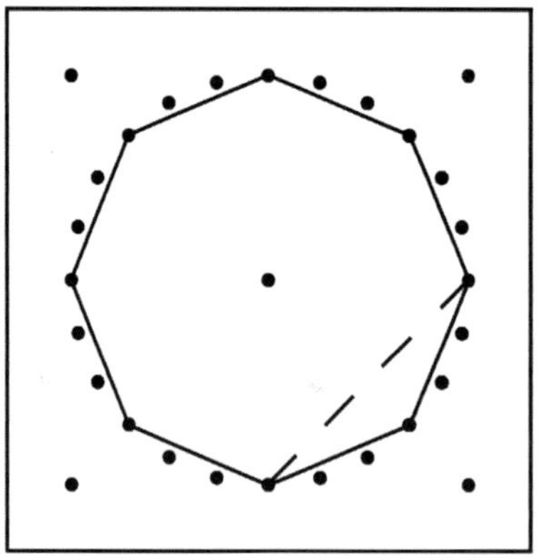

Draw an isosceles right triangle. Use the dashed line.

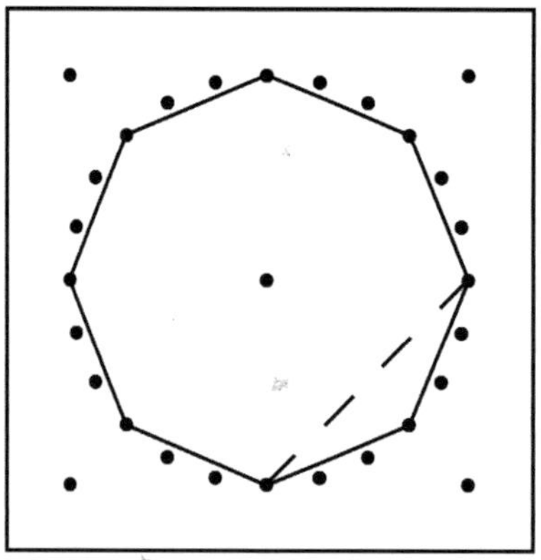

Draw an acute isosceles triangle. Use the dashed line.

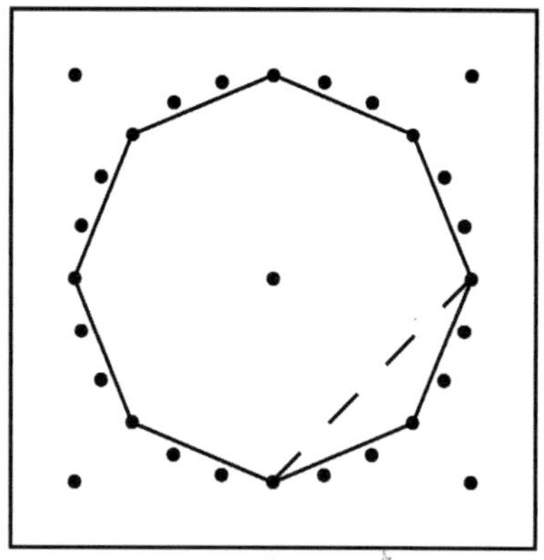

Draw an obtuse triangle. Use the dashed line.

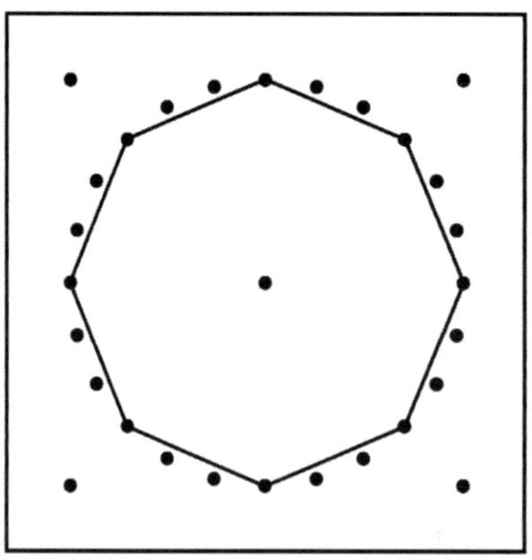

Make a trapezoid by drawing one line.

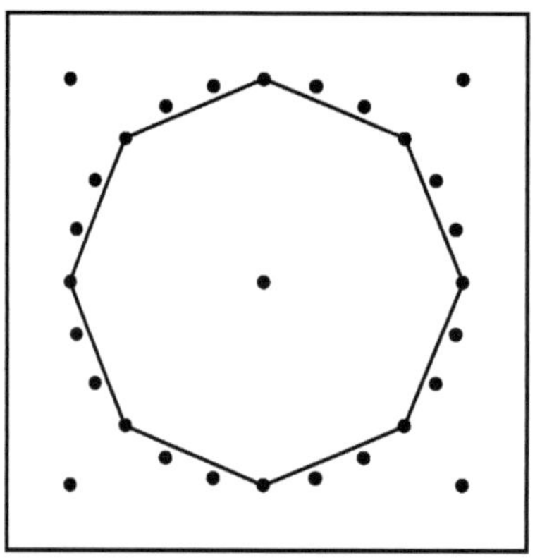

Make a trapezoid by drawing two lines.

How many of the new shapes have only one set of parallel lines? _______

How many of the new shapes have two sets of parallel lines? _______

How many of the new shapes have perpendicular lines? _______

Name: ______________________________

Date: ______________________________

Warm-Up

Multiply.

```
  8 9 6 ( )          5 8 9 6 ( )
×   7 3 ( )              × 8 4 ( )
_______________      _______________
          ( )                    ( )
```

Draw the lines of symmetry in the following regular polygons. Draw them freehand or with a straightedge.

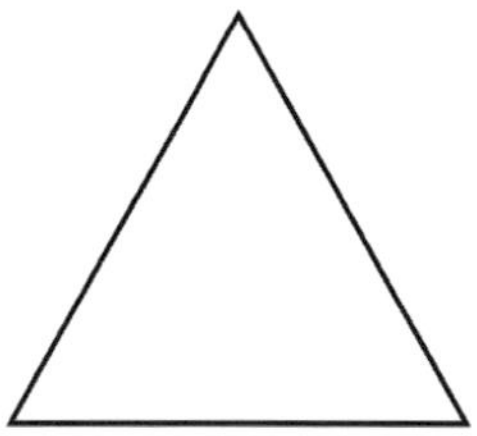

Lines of symmetry? _____

Lines of symmetry? _____

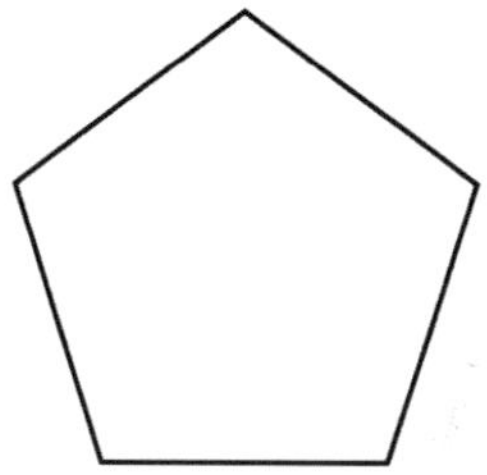

Lines of symmetry? _____

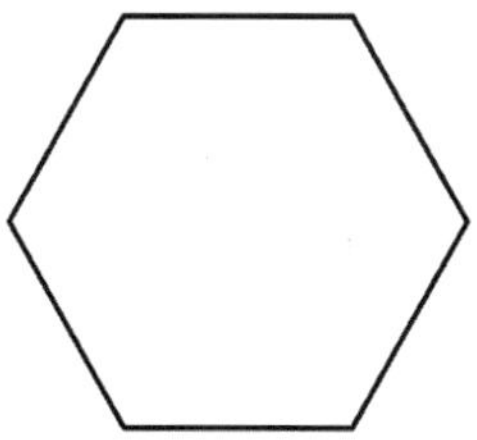

Lines of symmetry? _____

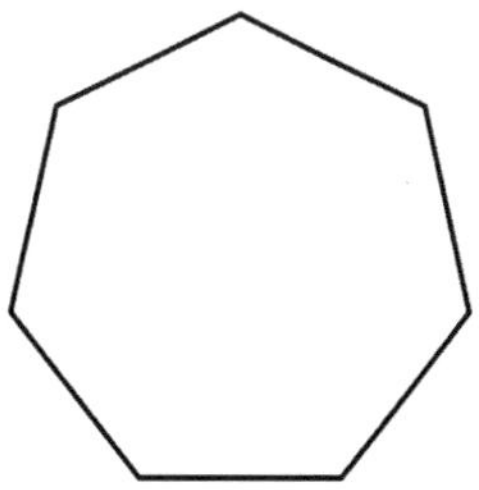

Lines of symmetry? _____

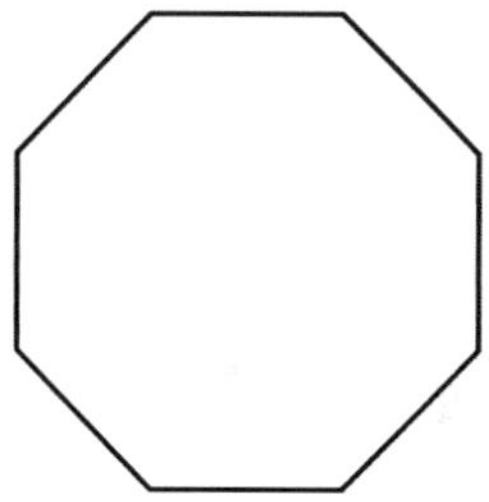

Lines of symmetry? _____

If you drew a regular polygon with 12 sides, how many lines of symmetry would it have? _______

If you drew a regular polygon with 100 angles, how many lines of symmetry would it have? _______

If the number of sides of a regular polygon is *n*, how many lines of symmetry does it have? _______

Draw the lines of symmetry in the following figures.

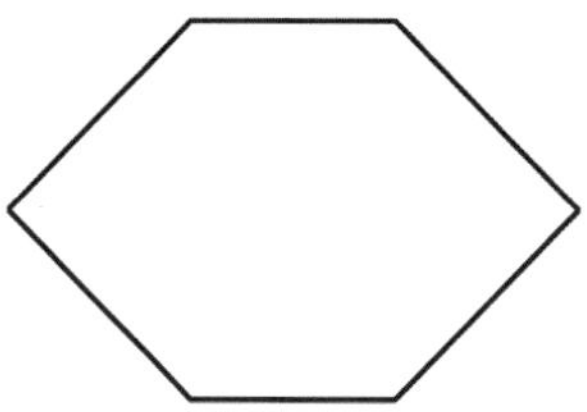

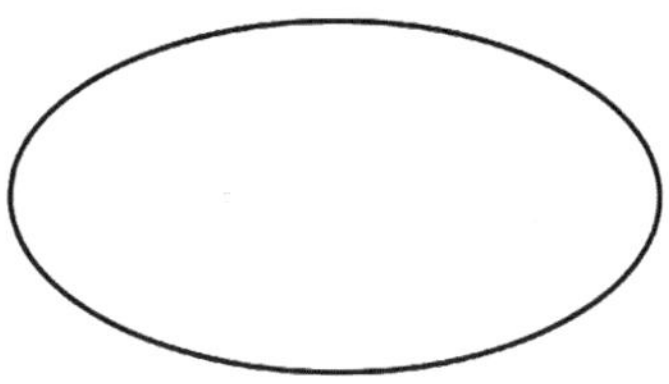

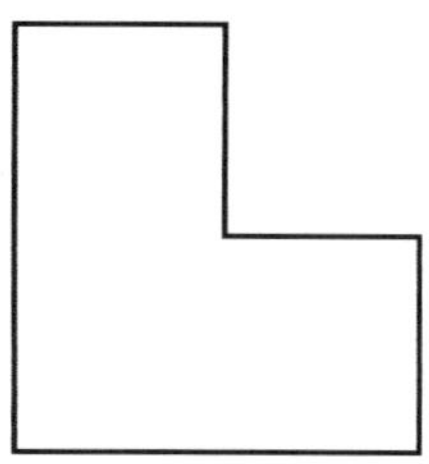

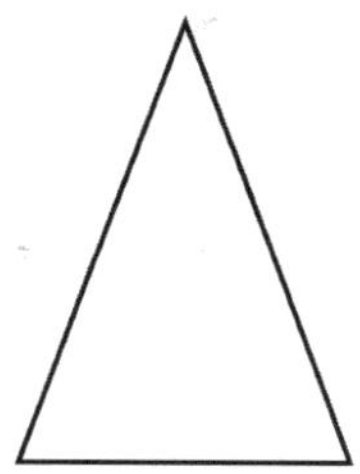

Name: ____________________ Date: ____________________

Reflect the shapes horizontally and vertically at the dashed lines. Use your drawing tools.

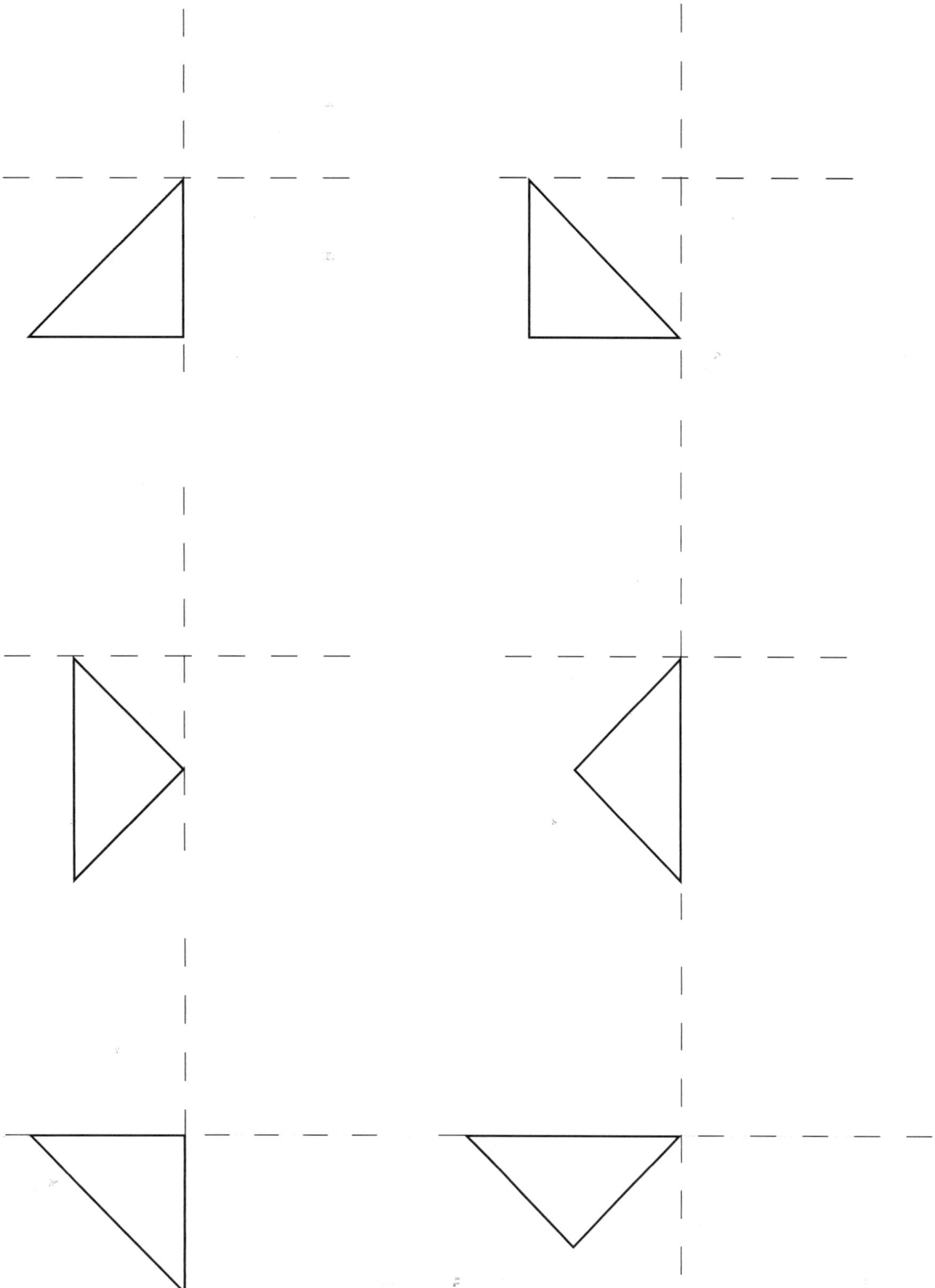

Name: ______________________ Date: ______________________

Reflect the design horizontally and vertically at the dashed lines. Use your drawing tools. No measuring or guessing.

1.

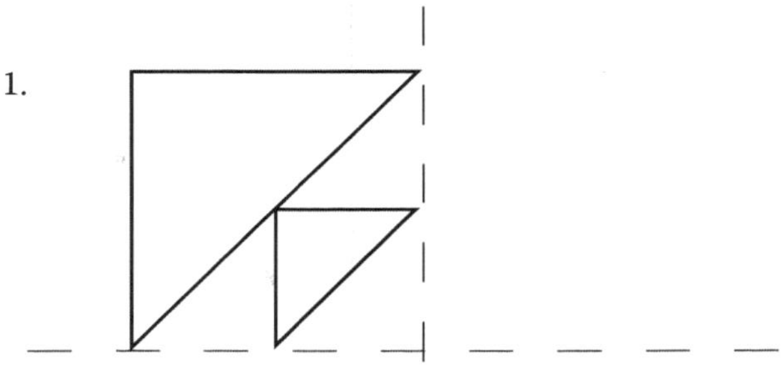

2.

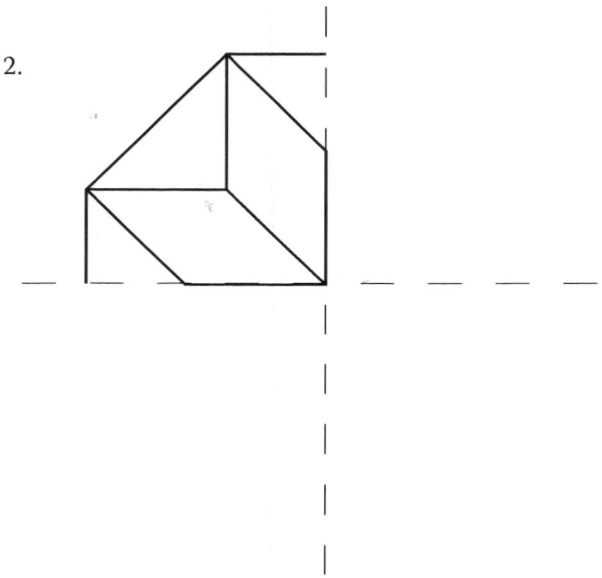

3. Both the 45 triangle and the 30-60 triangle are needed for this design.

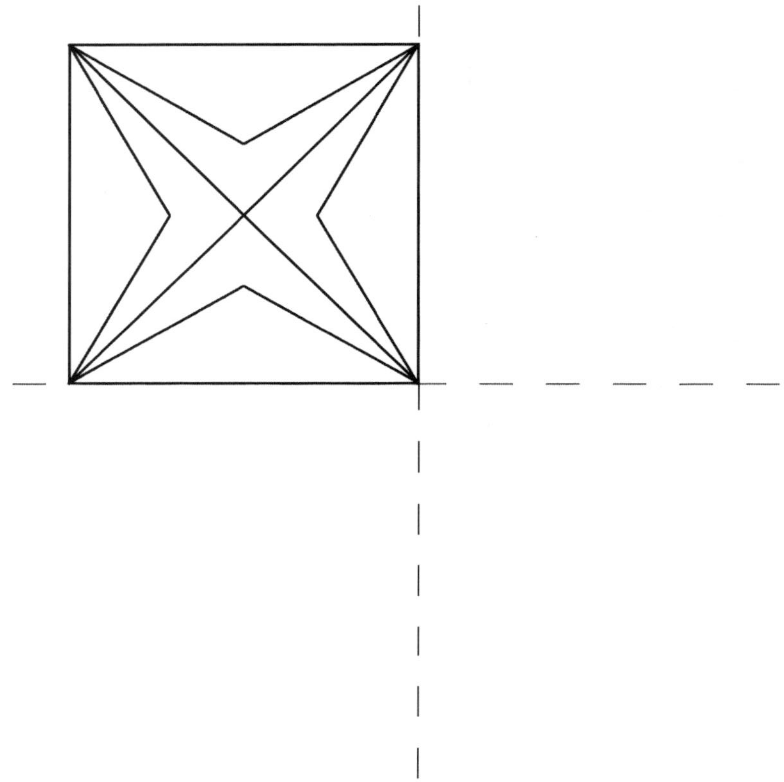

This design is from the interior of a cathedral in Uppsala, Sweden.

Name: ______________________________

Date: ______________________________

1. Make the left net below with the geometry panels. Then fold it up to make a cube, using the "B" panel as the bottom. Label the remaining faces of the panels using the following abbreviations:

B = Bottom T = Top F = Front A = Aft (back) L = Left R = Right

Repeat for the other two nets. Does it matter which panel is the bottom? ______

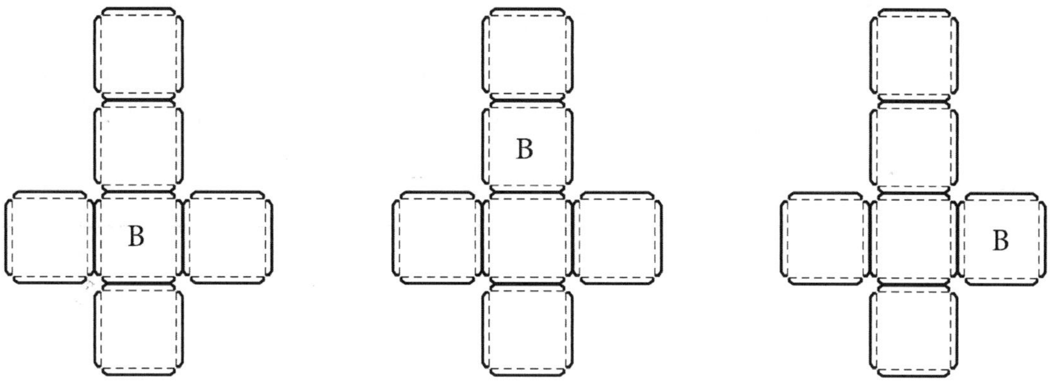

2. Make the following net with the geometry panels. Then fold it up to make a cube. Label the faces. Explain what happens.

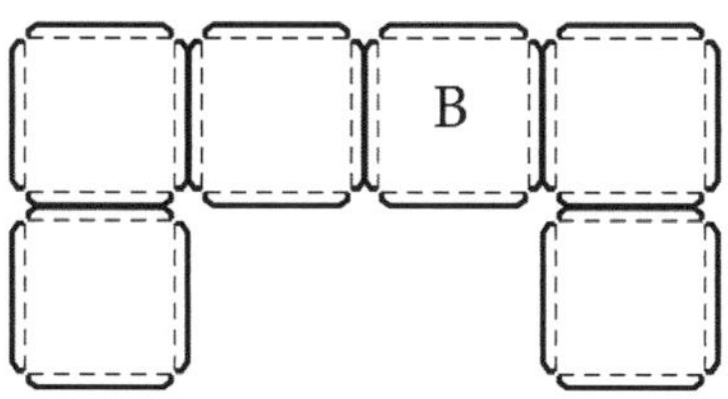

____

____

3. Fold each net below into a cube either mentally or with the geometry panels, using the "B" panel as the bottom. Then label the remaining faces. Write "yes" if it makes a cube or "no" if it does not.

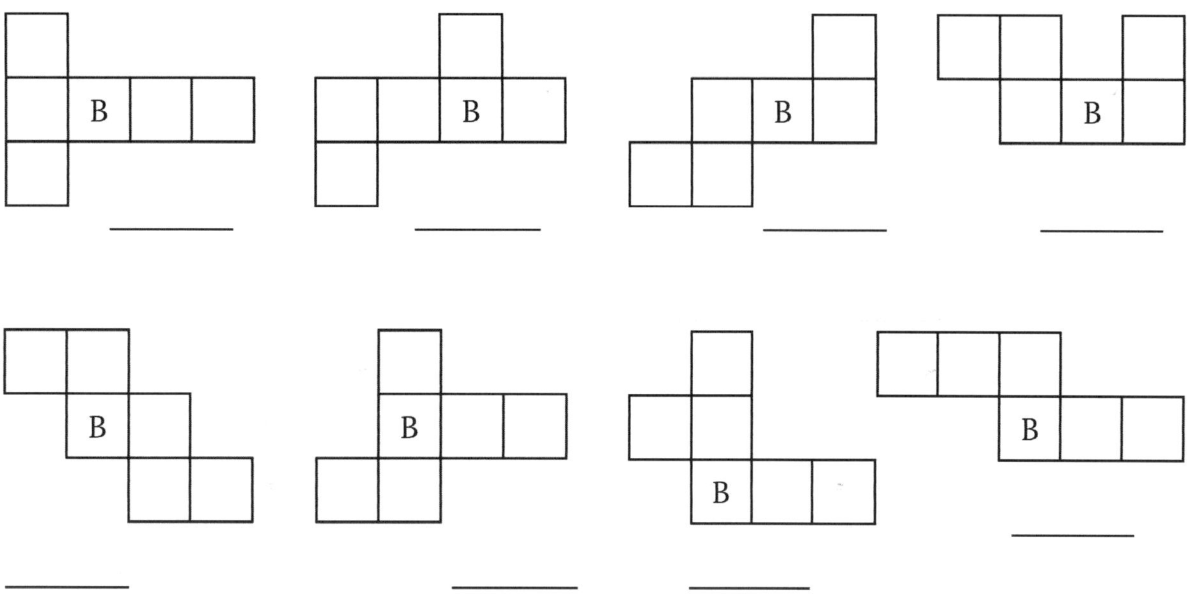

Name: ______________________________

Date: ______________________________

1. The drawing at the right is called an isometric drawing. Isometric drawings show equal distances along the three dimensions: width, length, and height. Engineers, architects, and designers use isometric drawings to show how a product will look.

The cube shown is 2 units on each edge. Using your drawing tools, draw another cube that is 3 units on an edge. Start at the dot. Make each side different by shading or crosshatching.

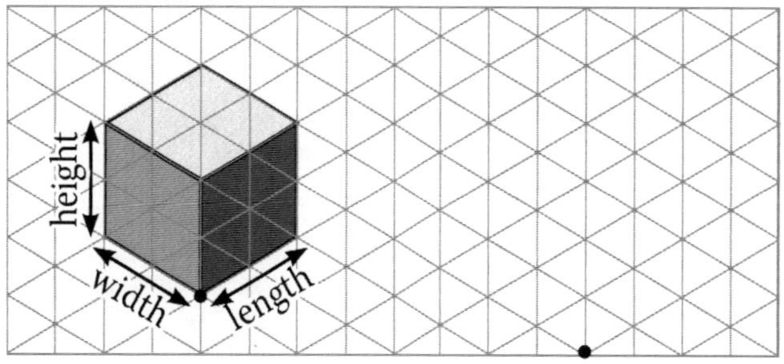

2. Copy the stairs, but make it one step higher.

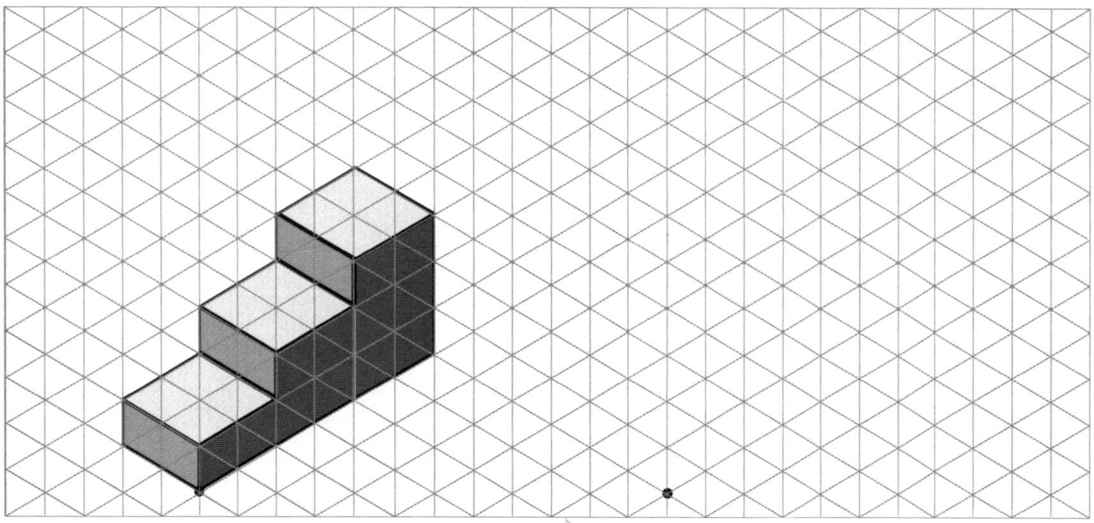

3. Draw a rectangular prism (a box) that is 2 units wide, 4 units long, and 3 units high.

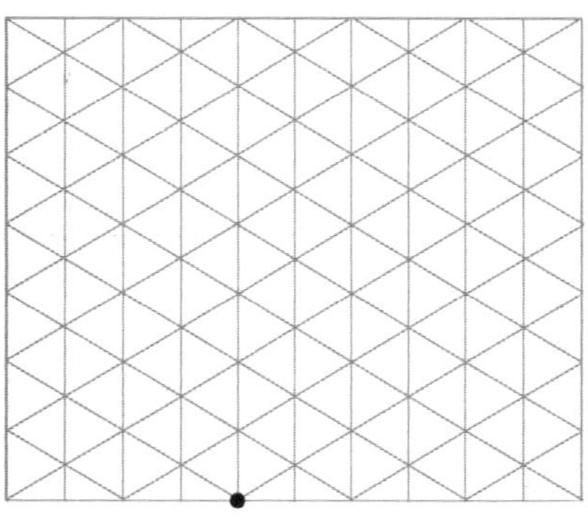

4. Draw an empty box like the box shown, but make it only 1 unit high. Be sure the bottom of the box shows.

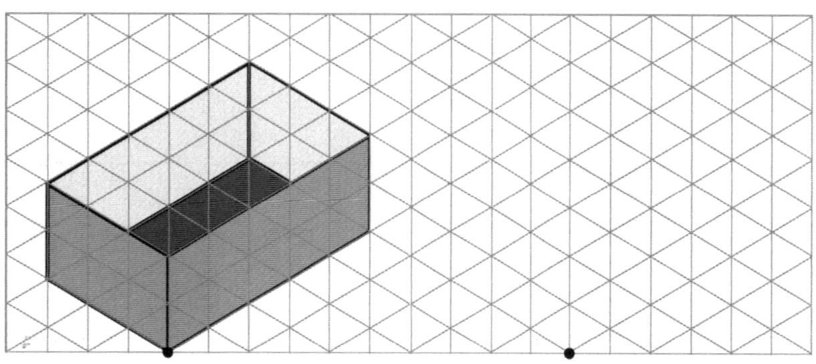

Name: ______________________

Date: ______________________

1. The drawing at the right is called an orthographic projection. Engineers, designers, and architects use these drawings to show exactly how objects look. The drawing shows the top, front, and side views of the 3-stair figure shown in the isometric drawing below. The small dots indicate the same point in all of the views.

Color or shade the front view in both figures the same way. Repeat for the top and side views.

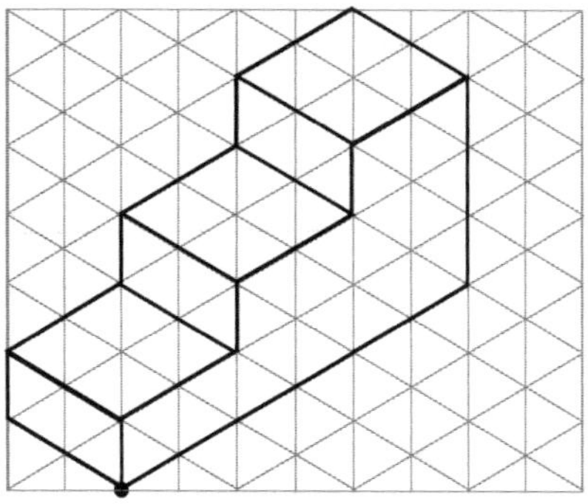

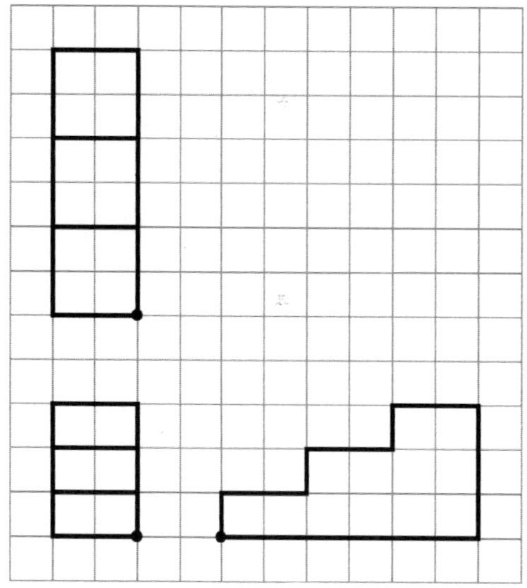

2. Draw the top, front, and side views of the 4-stair figure that you constructed in the last lesson using your drawing tools. First make the stairs with tiles.

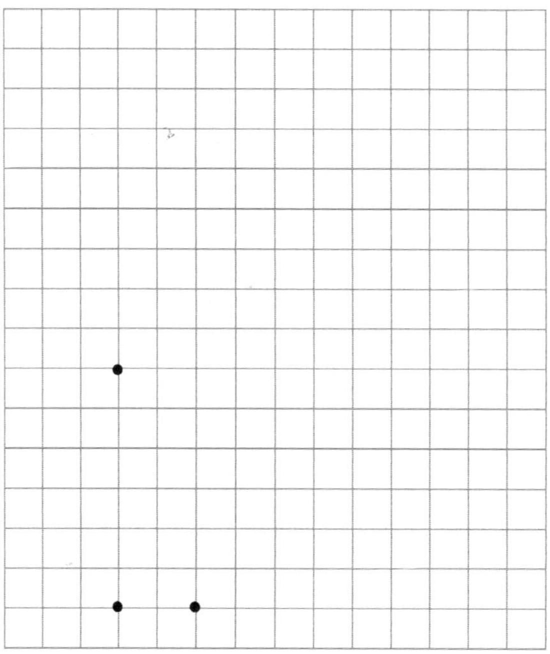

3–4. Draw the top, front, and side views of the rectangular prism and box that you constructed in the last lesson.

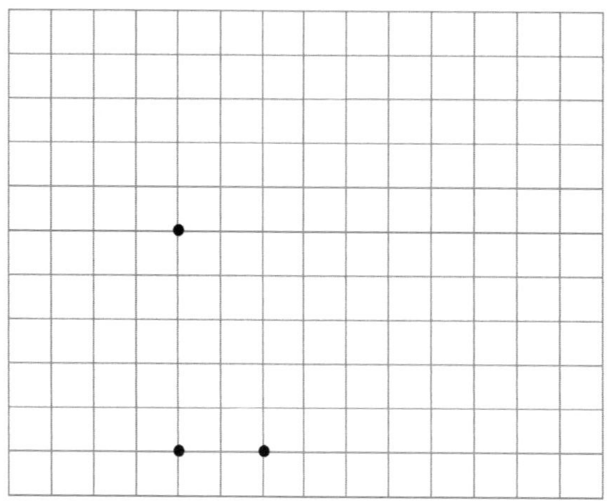

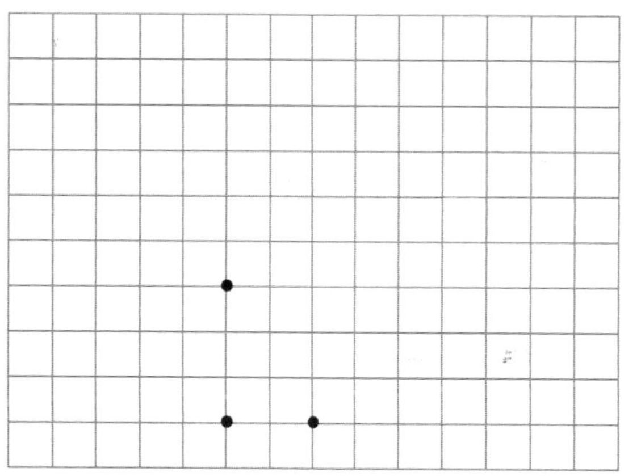

Name: ______________________________

Date: ______________________________

1. Draw the three views, top, front, and side, for a square pyramid that is 6 units high. The square base is 4 units on a side.

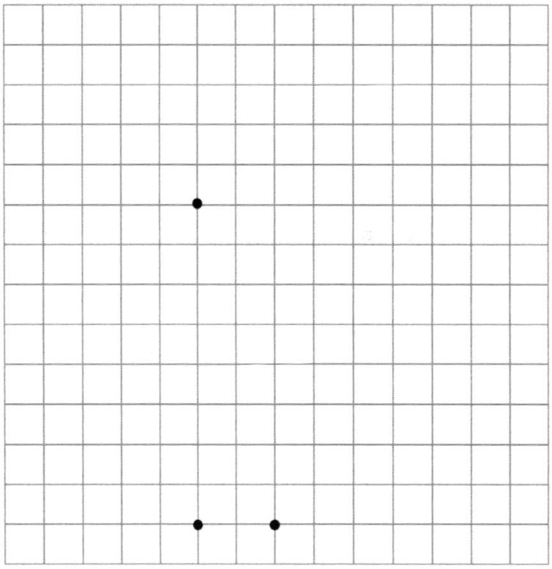

2. Draw the three views for the Problem 1 pyramid but now it is truncated so it is only 3 units high.

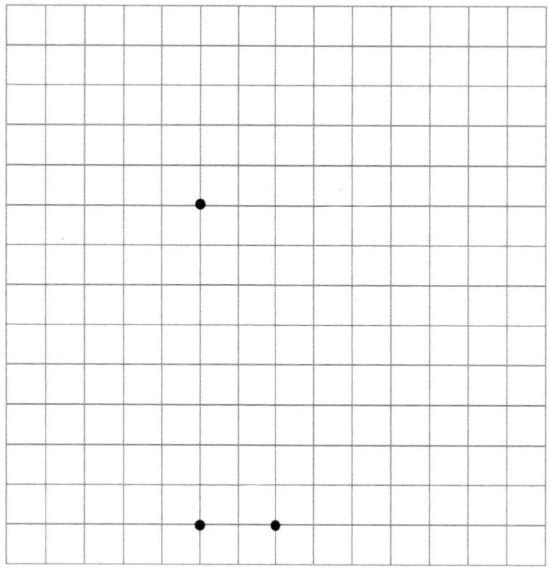

3. Draw the three views for a cone that is 8 units tall. The diameter of the base is 4 units. The "×" marks the center of the circle.

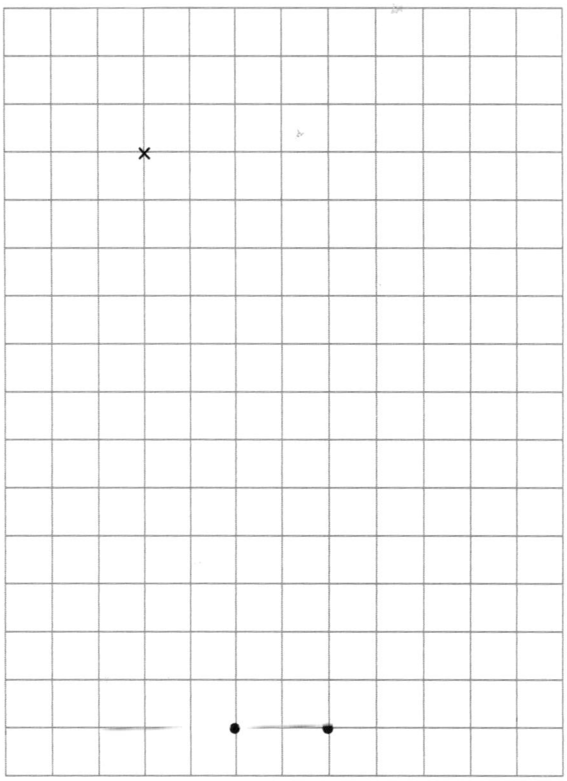

4. Draw the three views for the cone in Problem 3 but now it is truncated so it is now only 4 units tall. The "×" marks the center of the circle.

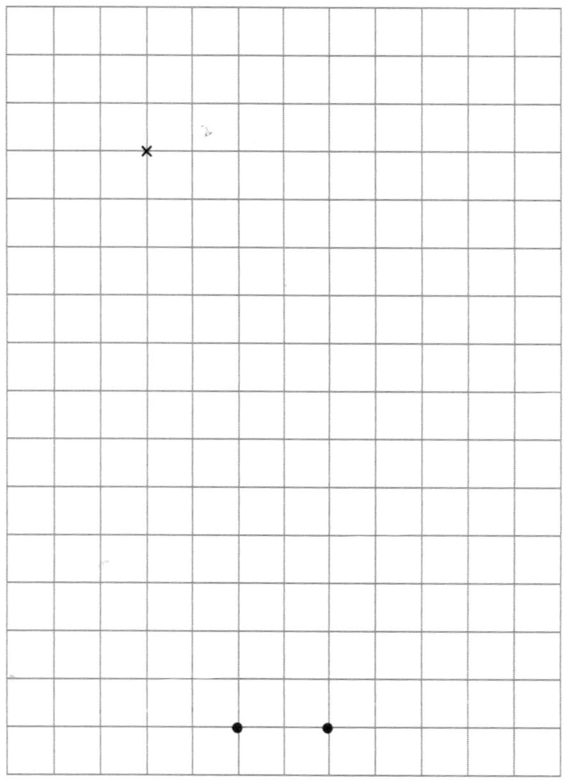

Worksheet 97, Name the Solids from Views

Name: ______________________________

Date: ______________________________

Name the geometric solids shown by the front, top, and side views.

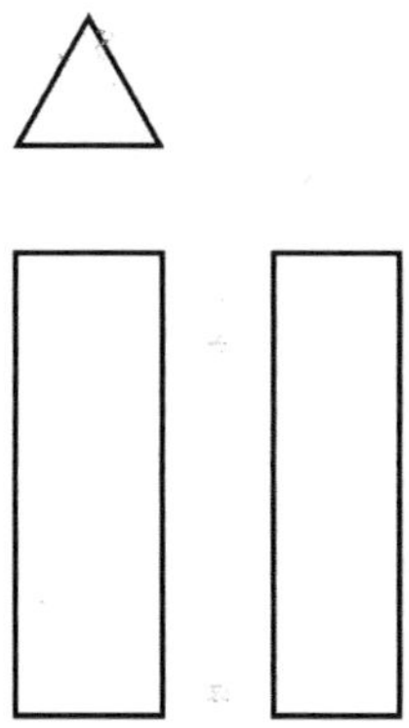

1. ____________________

____

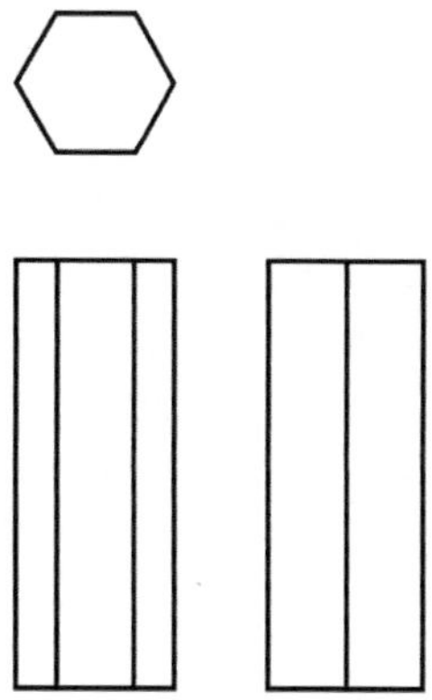

2. ____________________

____

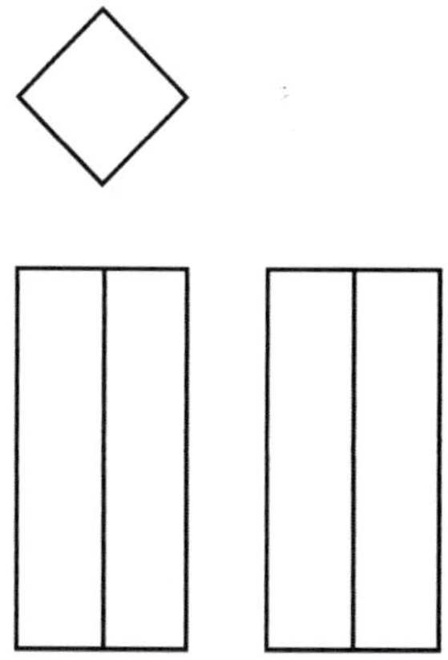

3. ____________________

____

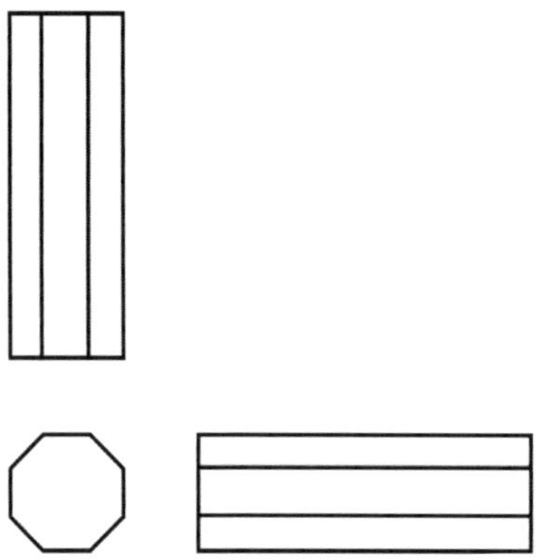

4. ____________________

____

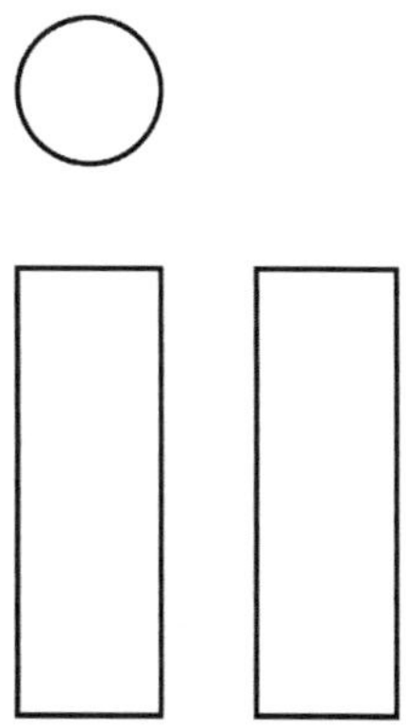

5. ____________________

____

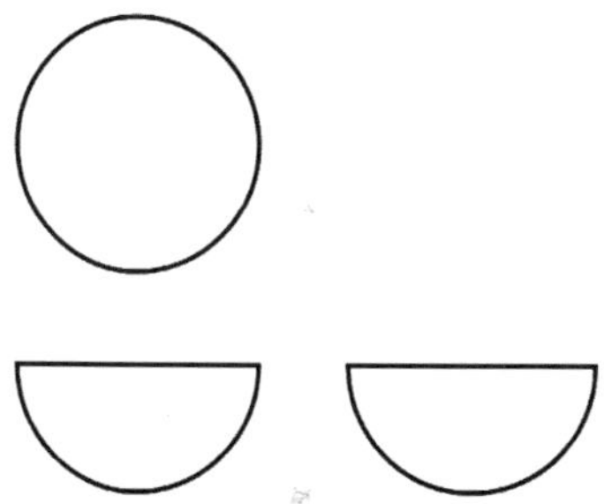

6. ____________________

____

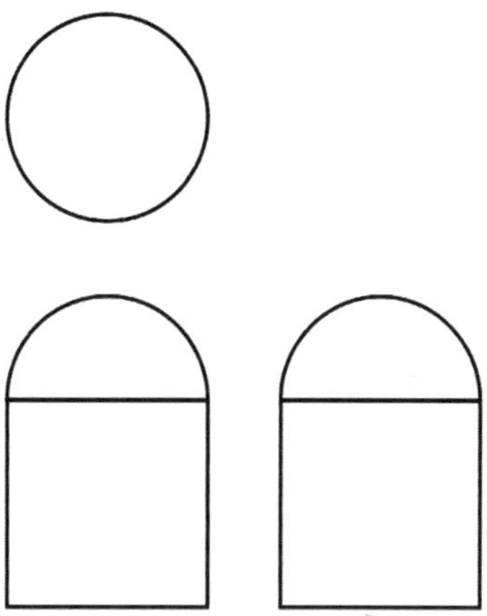

7. ____________________

____

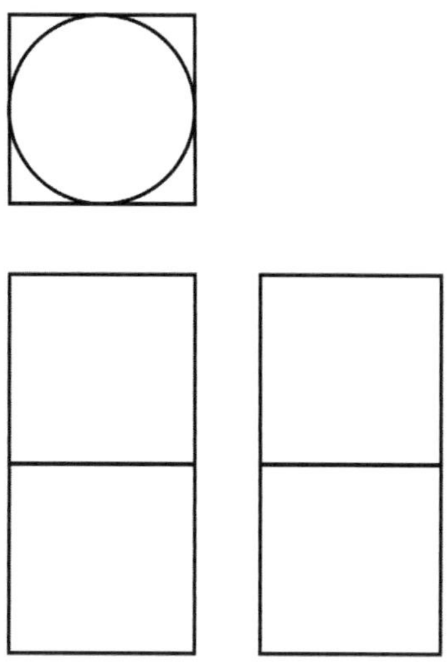

8. ____________________

____

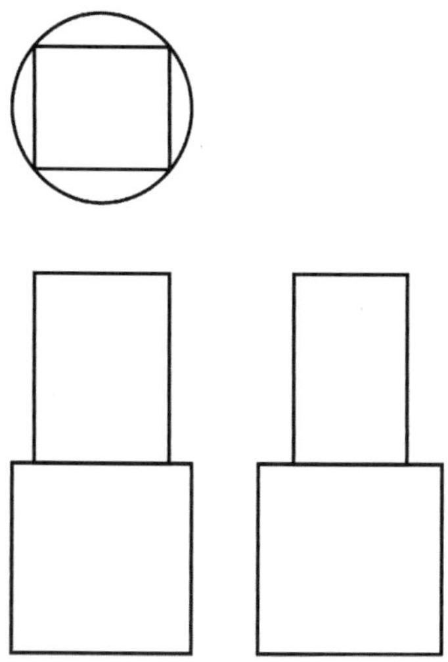

9. ____________________

____

Name: ______________________________

Date: ______________________________

INFORMATION: The *radius* of a circle is the distance from the center to a point on the perimeter. The word *radius* comes from the same word as *ray*, like the rays of the sun. A ray is a line that starts at a point and extends out. The plural of radius is *radii* or *radiuses*.

1. Make about 20 to 30 points each 1 in. away from the larger point below in all directions.

2. Draw a circle with a radius of 1 in. Draw a radius. What is the diameter ? ________

3. Draw four circles with the same center. Use radii of 1 in., 1.5 in., 2 in., and 2.5 in. These circles are called *concentric* circles.

Name: ______________________________

Date: ______________________________

4. Make the following design at the right. Use 3" as the radius for the large circle and all the arcs. It is sometimes called a "daisy."

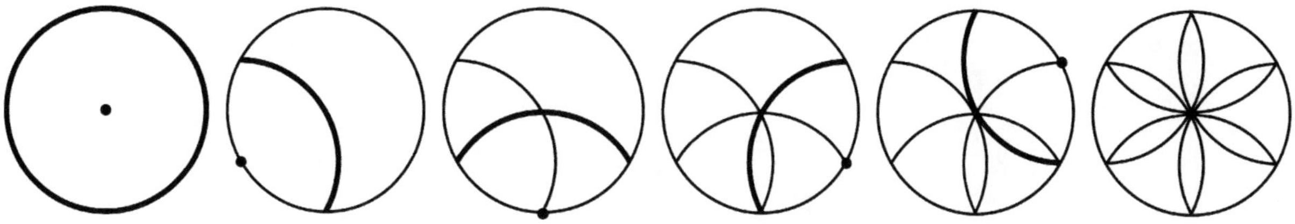

Name: ____________________________

Date: ____________________________

5. Choose a design below and draw it or make your own design.

Name: ______________________ Date: ______________________

On the right is a model of the Olympic rings. Draw the rings three times greater than the model on the right.

Hint: first locate the centers of all the rings; they will be three times farther apart than the model. The center of the first circle is given.

The colors of the rings, from left to right, are blue, yellow, black, green, and red. At least one of these colors is on the flag of every nation. Color the finished rings if you like.

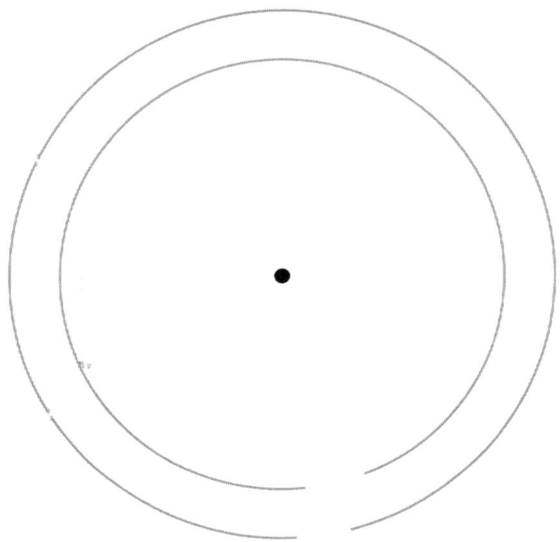

Worksheet 100, Area on the Geoboard

Name: ______________________________

Date: ______________________________

A square formed by four pegs on the geoboard is 1 unit of area.

Boundary points are pegs on the perimeter of the figure. A *boundary pair* is two boundary points.

Fill in the table for each figure below.

Problems 1 to 5.

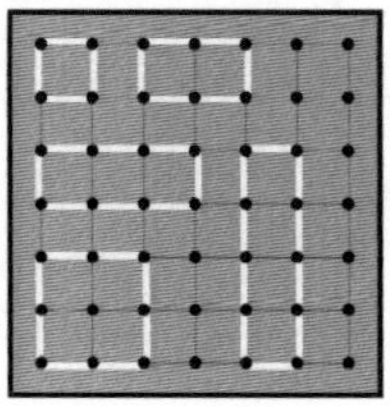

Area in Units	Number of Pegs	
	Boundary Pairs	Inside
1	2	

Problems 6 to 8.

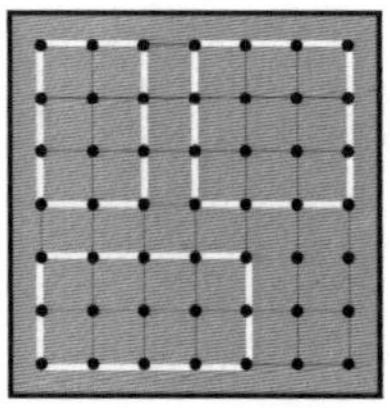

Area in Units	Number of Pegs	
	Boundary Pairs	Inside

Problems 9 to 13.

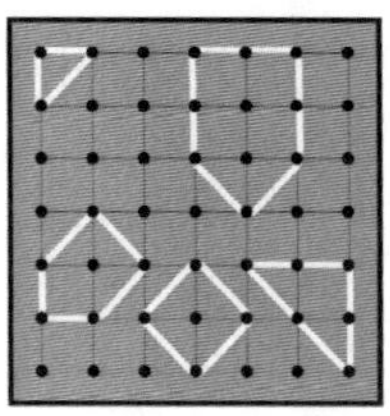

Area in Units	Number of Pegs	
	Boundary Pairs – 1	Inside

Name: ______________________________

Date: ______________________________

A square formed by four pegs on the geoboard has an area of 1 unit. Fill in the tables to find the area for each figure. Verify your answer with Pick's Theorem.

Problems 1 to 4.

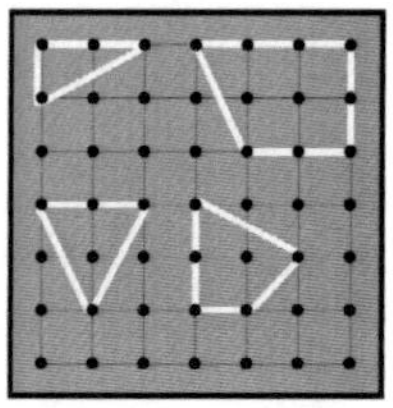

Problems 5 to 7.

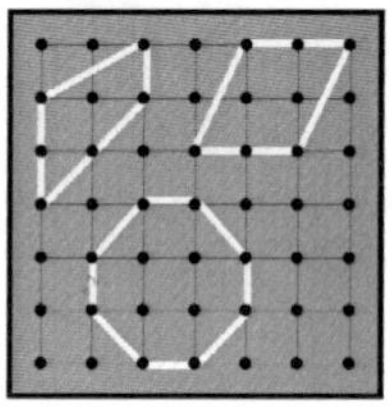

Area from Whole Units	Area from Half of 1 Unit	Area from Half of 2 Units	Total Area	Area using Pick's Theorem

Problems 8 to 11.

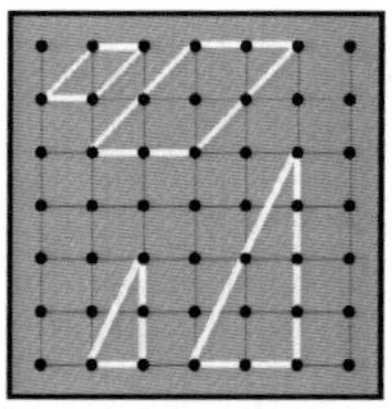

Problems 12 to 15.

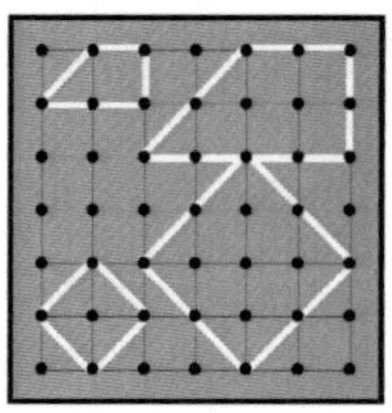

Area from Whole Units	Area from Half of 1 Unit	Area from Half of 2 Units	Total Area	Area using Pick's Theorem

What happens when the width and height of a figure is doubled?

____

Figures 16 to 19.

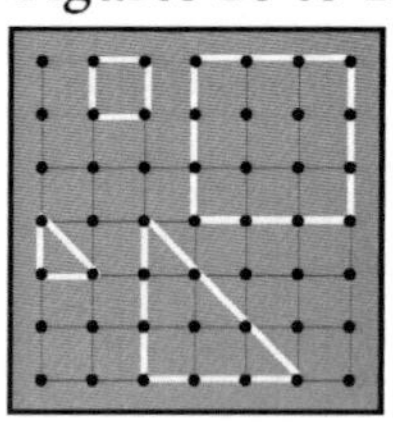

Area from Whole Units	Area from Half of 1 Unit	Area from Half of 2 Units	Total Area	Area using Pick's Theorem

What happens when the width and height of a figure is 3 times greater?

____

Name: ______________________

Date: ______________________

A square formed by four pegs on the geoboard is 1 unit of area.

Problems 1 to 5.

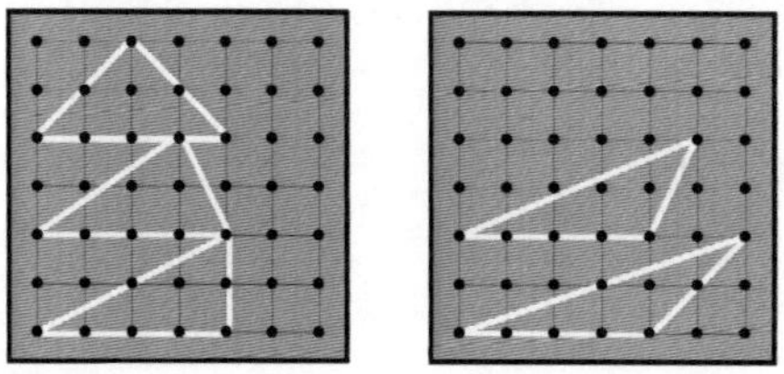

Find the area of each triangle, using any method.

1. _______ 2. _______ 3. _______ 4. _______ 5. _______

What is the base of each triangle? _______

What is the height of each triangle? _______

Problems 6 to 10.

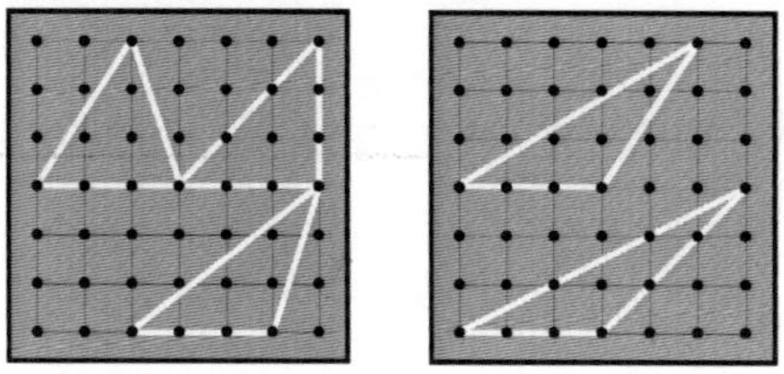

Find the area of each triangle, using any method.

6. _______ 7. _______ 8. _______ 9. _______ 10. _______

What is the base of each triangle? _______

What is the height of each triangle? _______

Problems 11 to 16.

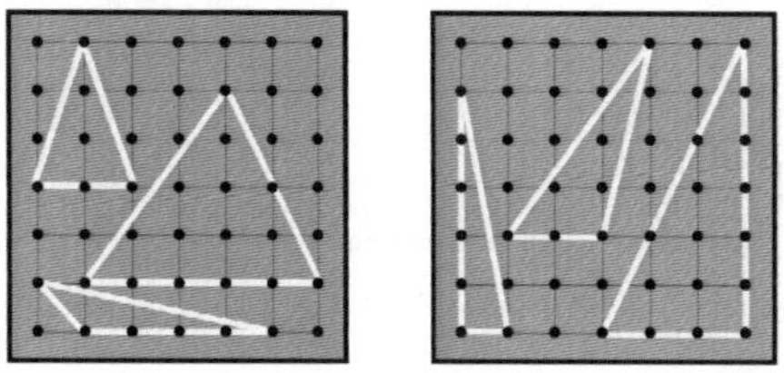

For each triangle, find the b (base), h (height), and area.

b	h	$b \times h$	Area

How is the area of a triangle related to $b \times h$?

____

Problems 17 to 20.

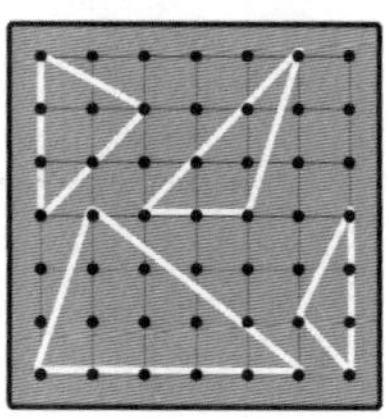

For each triangle, find the area by taking half of $b \times h$. Check your answer by another method.

b	h	$b \times h$	Area

Name: ______________________________

Date: ______________________________

1. How many squares with sides parallel to the edges of the geoboard can you make on the geoboard shown below? There are several different sizes of squares. Look for patterns.

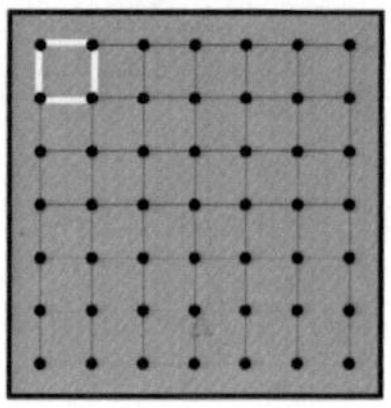

Size of square	Number in row	Number in column	Total

2. How many squares with sides 45° to the edges of the geoboard can you make?

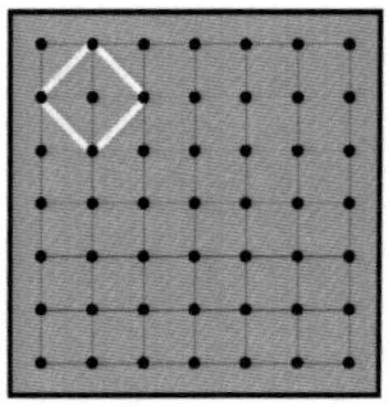

Size of square	Number in row	Number in column	Total

3. How many squares can you make on the circle side of the geoboard?

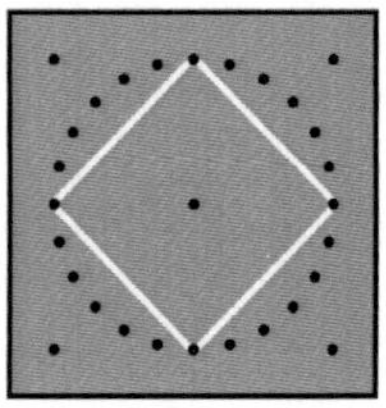

Describe squares	Total

4. How many rubber bands would you need to make all these squares at the same time? _______

5. How many different sizes of squares did you make on the geoboard for this lesson? _______

Name: ____________________

Date: ____________________

Warm-Up

Draw lines to match each triangle by sides and angles.

Scalene triangle		Right triangle
Equilateral triangle		Obtuse triangle
Isosceles triangle		Acute triangle

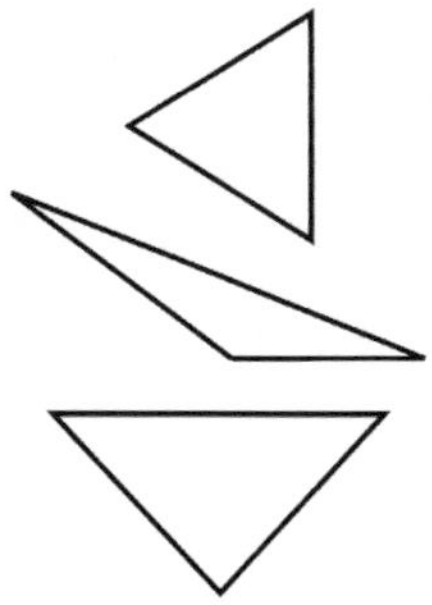

A square formed by four pegs on the geoboard is 1 unit of area. Connect the midpoints in the triangles below. Find the areas of of the large and small triangles, the parallelograms, and the trapezoids.

1.
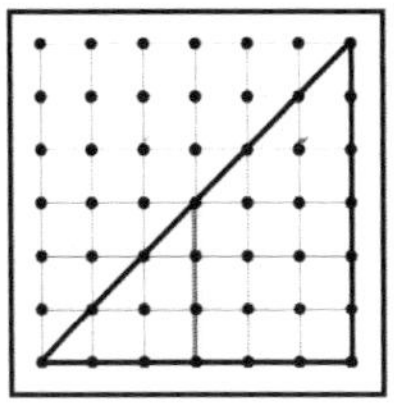

2.
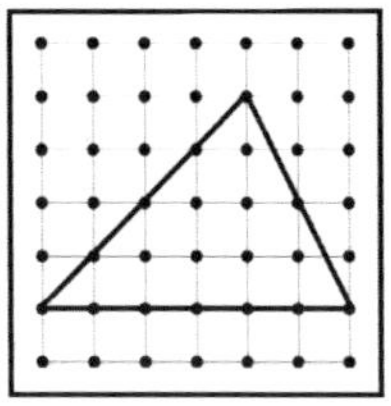

3.
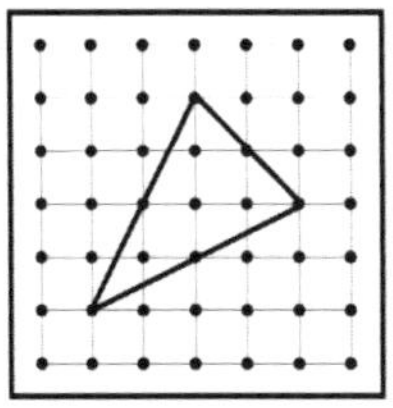

4.
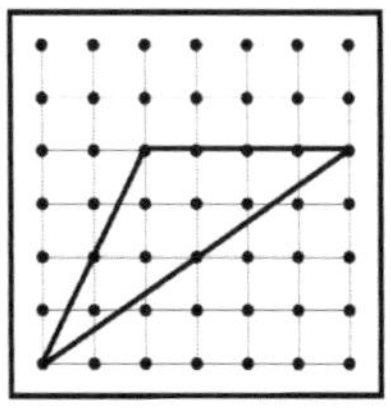

5.
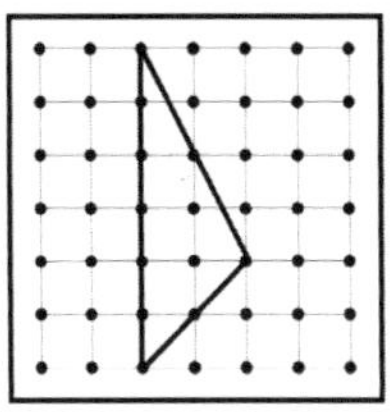

6.
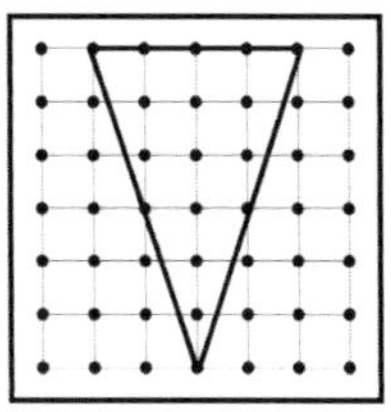

Area of large triangle	Area of small triangle	Area of parallelogram	Area of trapezoid

7. The area of the small triangles is what fraction of the area of the large triangles? _________
8. Are the four small triangles in the large triangle congruent? _________
9. Compare the length of a side of a small triangle to the opposite side of the large triangle? _________
10. The side of a small triangle is parallel or perpendicular to the opposite side of the large triangle? _____________
11. How many parallelograms are in each large triangle? _________
12. Compare the area of a parallelogram to the area of the large triangle? _________
13. How many trapezoids are in each large triangle? _________
14. Compare the area of a trapezoid to the area of the large triangle? _________

Worksheet 105, Midpoints in Quadrilaterals

Name: ______________________________

Date: ______________________________

A square formed by four pegs on the geoboard is 1 unit of area.

1. Connect the adjacent midpoints in the quadrilaterals below. Find the areas of of the quadrilaterals and the inside figures.

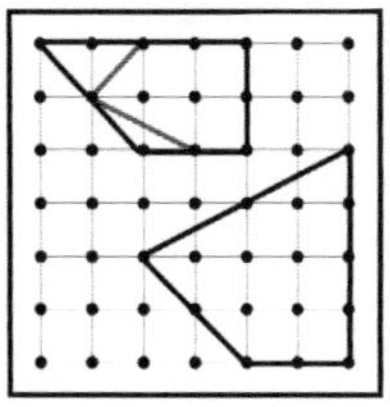

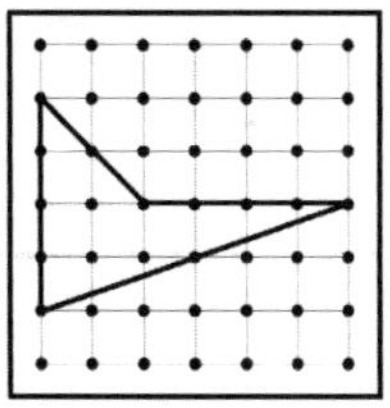

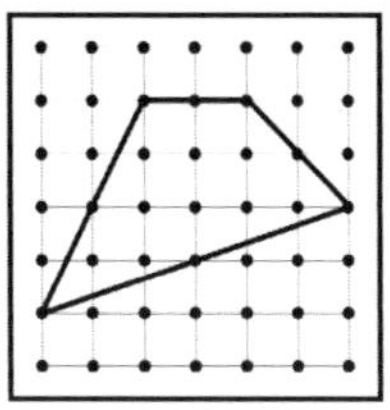

Area of quadrilateral	Area of inside figure

What figure is formed by connecting the midpoints in a quadrilateral? ______________________

What is the relationship between the area of the quadrilateral and the area of the inside figure?

____

2. Use the centimeter scale on your ruler. Connect adjacent midpoints in the quadrilateral below. What figure is it? ______________________

Cut out the four remaining triangles and put them together to make a parallelogram. (Hint: The lines you drew will form the edges of the parallelogram.)

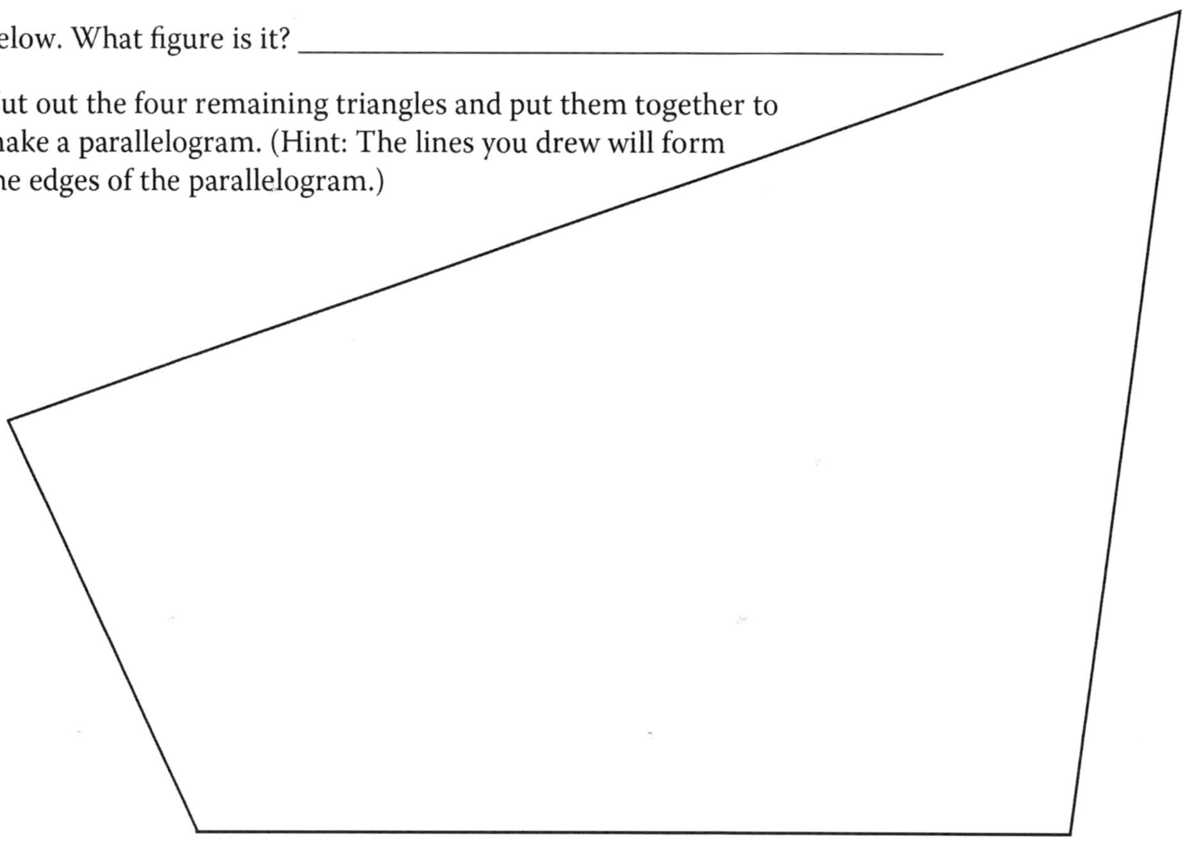

Name: ______________________________

Date: ______________________________

1-3. Write only the answers.

____

____

____

4-9. Write the answers.

718 + 265 = _____ 87 – 79 = _____

72 ÷ 9 = _____ 512 – 15 = _____

100 × 40 = _____ 200 – _____ = 68

10. Find 30,847 + 7946. 11. Find 8307 – 5457. 12. Find 6392 × 26. 13. Find 3450 ÷ 3.

14. Find 6729 ÷ 8.

15-16. Write these as numbers. Use commas to mark the periods.

27 million ____________________ 40 billion ____________________

17-22. Round the populations to the nearest million.

Australia	24,168,303	_____ million
Canada	36,134,016	_____ million
China	1,375,137,837	_____ million
Indonesia	259,281,096	_____ million
United Kingdom	63,742,977	_____ million
United States	324,308,763	_____ million

23. Write the countries in order of population, least to greatest.

24. The population of the world is 7,439,454,032. Write this number in words.

____

____

____

Name: ______________________________

Date: ______________________________

25-30. Prime numbers and factors.

How do you know that 6, 15, and 24 are not primes? ______________________

How do you know that 14, 28, and 63 are not primes? ______________________

Which of these numbers, 1, 13, 14, 25, 87, are prime numbers? ______________

Which of these numbers, 1, 13, 14, 25, 87, are composite? ______________

Name the factor pairs for 32. ______________________

What are the factors for 32. ______________________

31-34. Solve the equations.

$589 + 17 - 17 = n$

$n =$ _____

$314 + m = 815$

$m =$ _____

$(3 + p) \times 2 = 8$

$p =$ _____

$(q \times 4) + 2 = 30$

$q =$ _____

35-42. Write <, >, or = on the lines.

30×2 _____ 3×20

$13 \times 14 \times 15$ _____ $15 \times 14 \times 13$

$46 + 57$ _____ $46 + 59$

1×1 _____ $1 + 1$

2×2 _____ $2 + 2$

3×3 _____ $3 + 3$

$(8 \times 6) + (8 \times 7)$ _____ (8×13)

$73 - 59$ _____ $72 - 59$

43-45. Round 8748.

To nearest 10.	
To nearest 100.	
To nearest 1000.	

46-48. Round 1,583,804.

To nearest 1000.	
To nearest 10,000.	
To nearest 100,000.	

49-50. Solve the problem.

Bob worked from 11:00 a.m. to 2:00 p.m. for five days. He earned $7 an hour. Then he bought a gift for a friend for $25. How much money did he have left?

Worksheet 107, Fractions, Decimals, and Percents Review
Page 1

Name: ______________________

Date: ______________________

1-6. Write the non-shaded fractions on the lines.

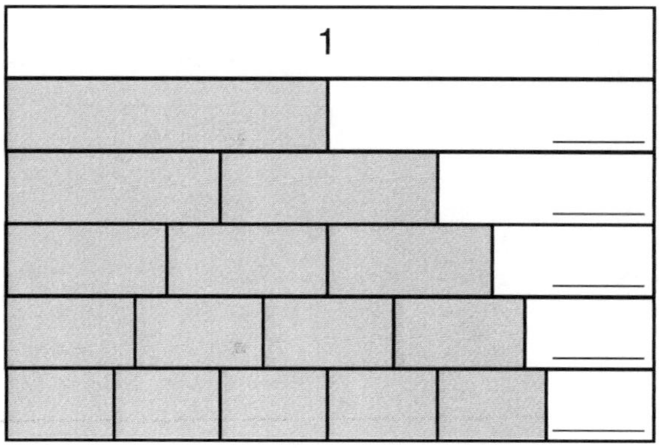

7-13. Circle True or False.

T F A fraction is always less than one.

T F A fraction is division with the numerator divided by the denominator.

T F The numerator tells the number of parts.

T F The denominator is the divisor.

T F If the numerator equals the denominator, the fraction equals zero.

T F A fraction is improper if the numerator is greater or equal to the denominator.

T F A mixed number is a whole number and a proper fraction.

14-18. Make a small mark at the correct position on the number line and write the fraction for the following: $1\frac{1}{4}$, $\frac{2}{3}$, $1\frac{1}{2}$, $\frac{1}{2}$, and $\frac{1}{4}$.

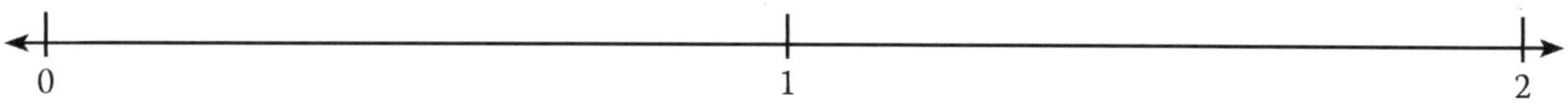

19-30. Fill in the blanks.

12 hours is what fraction of a day? _______

Four days is what fraction of a week? _______

20 minutes is what fraction of a hour? _______

50¢ is what fraction of a dollar? _______

11 inches is what fraction of a foot? _______

70 cm is what fraction of a meter? _______

One third of a day is _______ hours.

Two sevenths of a week is _______ days.

A quarter of a hour is _______ minutes.

Three quarters of a dollar is _______¢.

Two fourths of a foot _______ inches.

Four tenths of a meter is _______ cm.

31-34. Solve the problems.

An adult giraffe weighs about 900 kg. A newborn giraffe weighs one tenth of that. What does the baby weigh? How much weight does the baby gain before becoming an adult?

The perimeter of each regular figure below is 15 cm. What is the length of the sides of each figure?

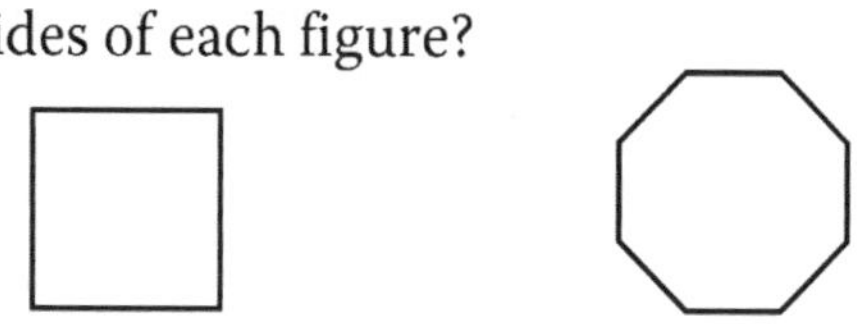

Name: ____________________

Date: ____________________

35-42. Match the expressions.

two fourths	6
$3 \div 2$	$\frac{1}{2}$
two thirds of 9	$\frac{2}{3}$
$\frac{1}{2}$ of $\frac{1}{4}$	$1\frac{1}{2}$
$\frac{1}{3} \times 2$	$\frac{1}{8}$
$\frac{4}{10} + \frac{9}{10}$	$1\frac{7}{10}$
$2 \div 9$	$1\frac{3}{10}$
$2 - \frac{3}{10}$	$\frac{2}{9}$

43-47. Write as decimals.

$\frac{4}{10}$ ________

$\frac{17}{100}$ ________

$\frac{9}{100}$ ________

$3\frac{7}{100}$ ________

$4\frac{50}{100}$ ________

48-52. Write as fractions.

0.7 ________

.06 ________

.87 ________

0.95 ________

2.05 ________

53-58. Put these fractions and decimals in order from least to greatest.

$\frac{5}{10}$	$\frac{5}{100}$	.04	____________________
$\frac{300}{100}$	.03	.33	____________________
.5	$\frac{49}{100}$	$\frac{51}{100}$	____________________
1.0	$\frac{9}{10}$	$\frac{1}{3}$	____________________
.06	$\frac{3}{4}$	$\frac{50}{100}$	____________________

59-65. Match the fractions and percentages.

$\frac{1}{4}$	50%
$\frac{7}{10}$	25%
$\frac{2}{4}$	70%
1	40%
$\frac{1}{3}$	100%
$\frac{2}{5}$	75%
$\frac{3}{4}$	$33\frac{1}{3}$%

66-67. Solve the problem.

Robin is trying to decide which gift to buy. The blue one is $24 and is 15% off. The green one is $26 and is 20% off. Which one is cheaper?

Worksheet 108, Geometry and Measurement Review
Page 1

Name: ______________________________

Date: ______________________________

1-10. Use these figures for the questions below.

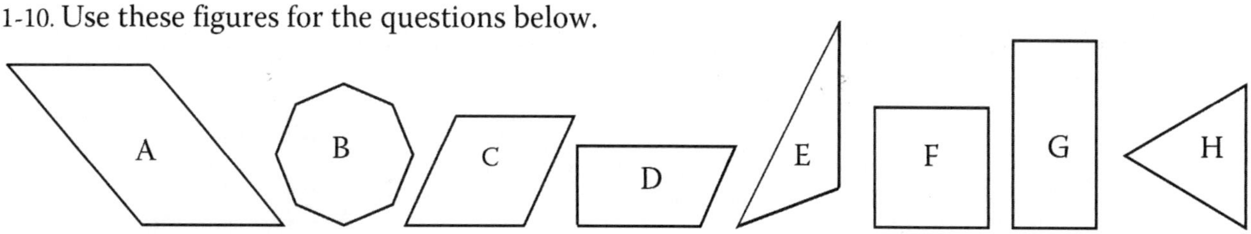

Which figures have parallel lines? ______________________

Which figures have perpendicular lines? ______________________

Which figures have all sides congruent? ______________________

Which figures are regular polygons? ______________________

Which figures are quadrilaterals? ______________________

Which figures are parallelograms? ______________________

Which figures are rhombuses? ______________________

Which figures are rectangles? ______________________

Which figures are trapezoids? ______________________

Which figures do not have any lines of symmetry? ________________

11-16. Match the angles with their measurement.

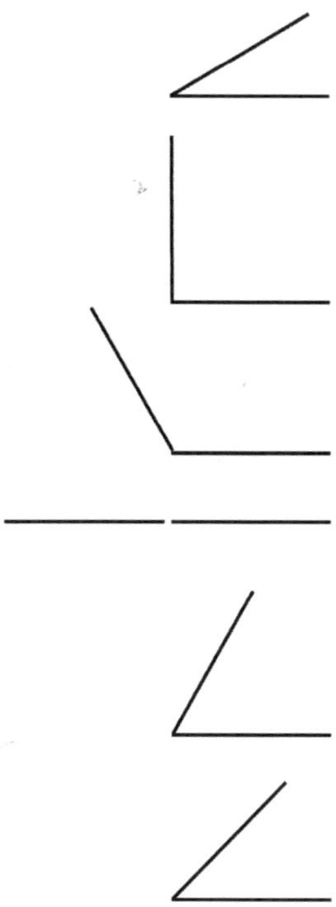

30°

45

60°

90°

120°

180°

17-18. Use a ruler to measure the triangle in tenths of a inch. Then find the perimeter.

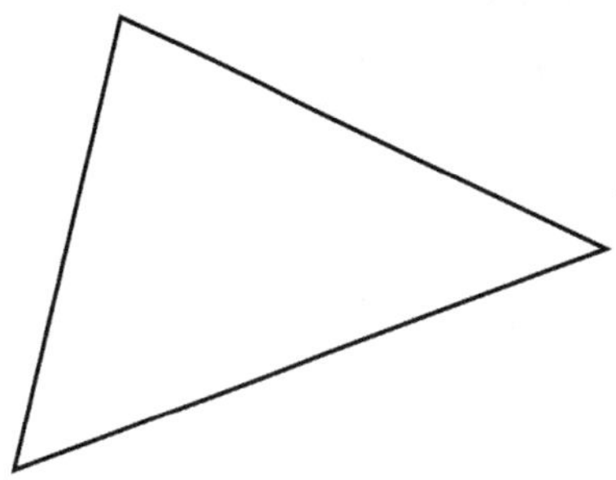

Name: ________________________

Date: ________________________

19-24. Draw lines to match each triangle by sides and angles.

Scalene triangle	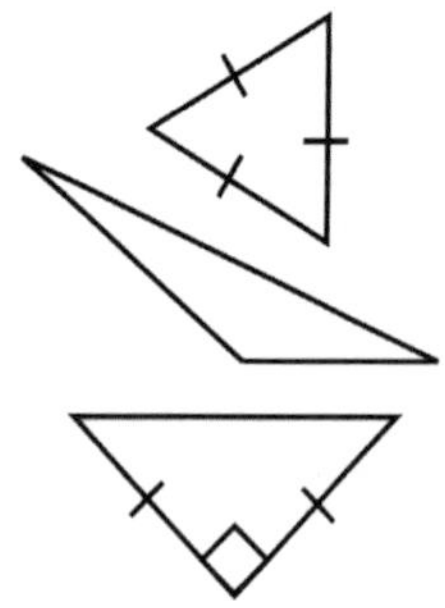	Right triangle
Equilateral triangle		Obtuse triangle
Isosceles triangle		Acute triangle

25-27. Draw all the lines of symmetry in the figures below and answer the questions.

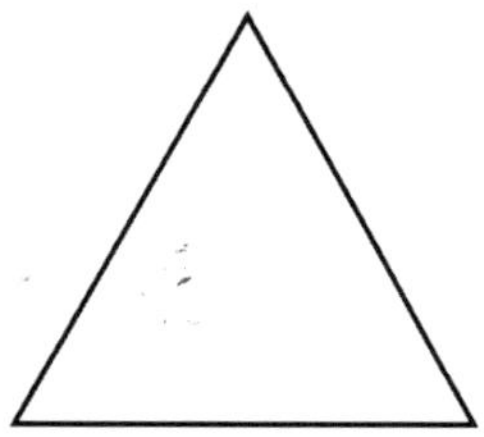

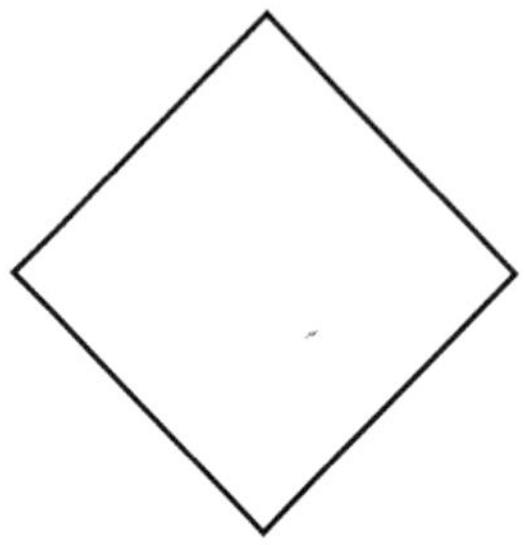

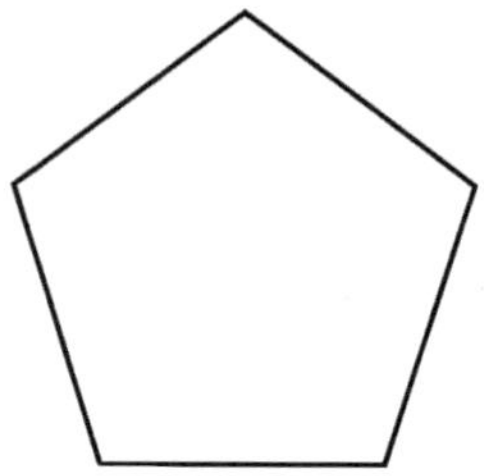

How many lines of symmetry?	How many lines of symmetry?	How many lines of symmetry?

28-41. Fill in the blanks.

How many degrees are in a circle? ________

Name the angles in an isosceles right triangle. ____________

Name the angles in an equilateral triangle. ____________

How many angles does an octagon have? _______

Is a square a rectangle? ________

Can a rhombus be a square? ________

Is a circle a polygon? ________

How many inches are in a foot? _______

How many feet are in a yard? _______

How many centimeters are in a decimeter? _______

How many decimeters are in a meter? _______

How many centimeters are in a meter? _______

How many milliliters are in a liter? _______

Which is longer, an inch or a centimeter? _______

Name: ______________________________

Date: ______________________________

1-3. Write only the answers.

____

____

____

4-9. Write the answers.

737 + 218 = _____

94 – 86 = _____

63 ÷ 7 = _____

407 – 15 = _____

100 × 60 = _____

200 – _____ = 57

10. Find 46,893 + 6827.

11. Find 6015 – 3837.

12. Find 7638 × 36.

13. Find 4328 ÷ 4.

14. Find 5847 ÷ 7.

15-16. Write these as numbers. Use commas to mark the periods.

103 million ______________________

56 billion ______________________

17-22. Round the populations to the nearest thousand.

Cairo, Egypt	9,278,441	_____ thousand
Istanbul, Turkey	13,710,512	_____ thousand
Lima, Peru	8,481,415	_____ thousand
Mumbai, India	12,478,447	_____ thousand
New York City, US	8,550,405	_____ thousand
Shanghai, China	18,626,300	_____ thousand

23. Write the cities in order of population, least to greatest.

24. The population of North America is 360,767,848. Write this number in words.

____

____

____

Name: ______________________________

Date: ______________________________

25-30. Prime numbers and factors.

How do you know that 15, 25, and 50 are not primes? ______________________

How do you know that 27, 36, and 81 are not primes? ______________________

Which of these numbers, 1, 2, 5, 32, 53, are prime numbers? ______________________

Which of these numbers, 1, 2, 5, 32, 53, are composite? ______________________

Name the factor pairs for 20. ______________________

What are the factors for 20. ______________________

31-34. Solve the equations.

$376 - 23 + 23 = n$

$n =$ _____

$297 + m = 308$

$m =$ _____

$(5 + p) \times 3 = 18$

$p =$ _____

$(q \times 5) + 1 = 26$

$q =$ _____

35-42. Write <, >, or = on the lines.

40×2 _____ 4×20

$13 \times 14 \times 15$ _____ $15 \times 14 \times 13 \times 0$

$85 + 57$ _____ $87 + 57$

$1 + 1$ _____ 1×1

$2 + 2$ _____ 2×2

$4 + 4$ _____ 4×4

$(7 \times 10) - (7 \times 7)$ _____ (7×3)

$37 - 19$ _____ $35 - 19$

43-45. Round 6489.

To nearest 10.	
To nearest 100.	
To nearest 1000.	

46-48. Round 3,694,758.

To nearest 1000.	
To nearest 10,000.	
To nearest 100,000.	

49-50. Solve the problem.

Rochelle rode her bike from 9:00 a.m. to 1:00 p.m. everyday for four days. She biked 6 miles in an hour. Then she biked an extra 3 miles. How far did she bike?

Name: ______________________

Date: ______________________

51-56. Write the non-shaded fractions on the lines.

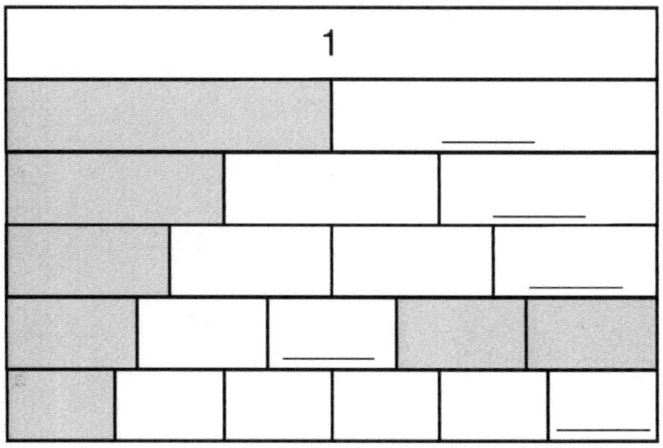

57-63. Circle True or False.

T F A fraction can never be greater than one.

T F A fraction is division with the denominator divided by the numerator.

T F The numerator tells the number of parts.

T F The denominator is the divisor.

T F If the numerator equals the denominator, the fraction equals one.

T F A fraction is proper if the numerator is more than the denominator.

T F A mixed number is a whole number and an improper fraction.

64-68. Make a small mark at the correct position on the number line and write the fraction for the following: $\frac{1}{3}$, $\frac{2}{3}$, $1\frac{1}{3}$, $\frac{1}{2}$, and $1\frac{3}{4}$.

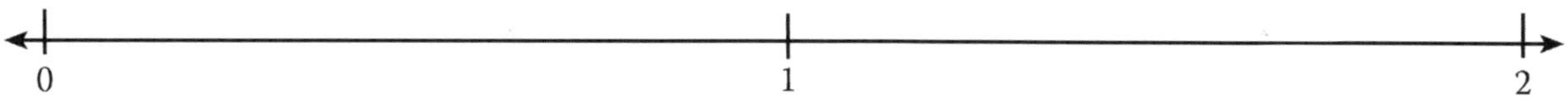

69-80. Fill in the blanks.

90 cm is what fraction of a meter? _______

Seven days is what fraction of April? _______

$1.25 is what fraction of a dollar? _______

8 hours is what fraction of a day? _______

15 minutes is what fraction of a hour? _______

5 inches is what fraction of a foot? _______

One half of a day is _______ hours.

One seventh of a week is _______ hours.

5 quarters of a hour is _______ minutes.

Three quarters of a dollar is _______¢.

Three fourths of a foot _______ inches.

One tenth of a meter is _______ cm.

81-84. Solve the problems.

An adult blue whale can be 32 ft long. At birth the baby is about one fourth that long. How long is the baby? How much longer does the baby whale grow before it is an adult?

The perimeter of each regular figure below is 13 cm. What is the length of the sides of each figure?

Name: ______________________________

Date: ______________________________

85-92. Match the expressions.

$4 \div 3$	4
$\frac{7}{10} + \frac{6}{10}$	$1\frac{3}{10}$
two fifths of 10	$\frac{1}{6}$
$5 \div 2$	$\frac{4}{3}$
$\frac{1}{2}$ of $\frac{1}{3}$	$2\frac{1}{2}$
$2 - \frac{3}{4}$	$\frac{2}{7}$
$\frac{1}{3} \times 3$	1
$2 \div 7$	$1\frac{1}{4}$

93-97. Write as decimals.

$\frac{3}{10}$ ________

$\frac{25}{100}$ ________

$\frac{9}{100}$ ________

$1\frac{59}{100}$ ________

$3\frac{8}{100}$ ________

98-102. Write as fractions.

0.83 ________

.4 ________

1.5 ________

0.01 ________

2.60 ________

103-108. Put these fractions and decimals in order from least to greatest.

$\frac{9}{10}$	$\frac{1}{9}$	1.0	____________________
$\frac{7}{10}$	$\frac{7}{100}$	.06	____________________
.51	$\frac{49}{100}$	$\frac{52}{100}$	____________________
.01	$\frac{5}{4}$	$\frac{50}{100}$	____________________
.66	$\frac{600}{100}$	.06	____________________

109-115. Match the fractions and percentages.

$\frac{3}{4}$	30%
$\frac{3}{10}$	$33\frac{1}{3}$%
$\frac{3}{2}$	75%
1	100%
$\frac{1}{3}$	25%
$\frac{2}{5}$	150%
$\frac{1}{4}$	40%

116-117. Solve the problem.

Terry is working on a fundraiser for charity and has collected \$84. Terry can keep 15%. How much does Terry earn? How much goes to the charity?

Name: ____________________

Date: ____________________

119-128. Use these figures for the questions below.

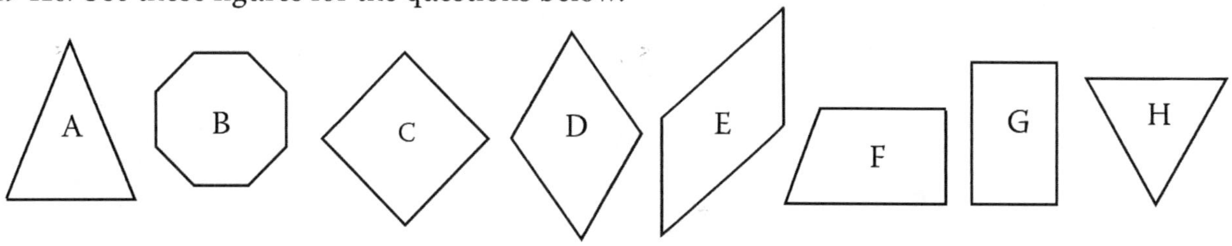

Which figures have parallel lines? ____________________

Which figures have perpendicular lines? ____________________

Which figures have all sides congruent? ____________________

Which figures are regular polygons? ____________________

Which figures are quadrilaterals? ____________________

Which figures are parallelograms? ____________________

Which figures are rhombuses? ____________________

Which figures are rectangles? ____________________

Which figures are trapezoids? ____________________

Which figures do not have any lines of symmetry? ____________________

129-134. Match the angles with their measurement.

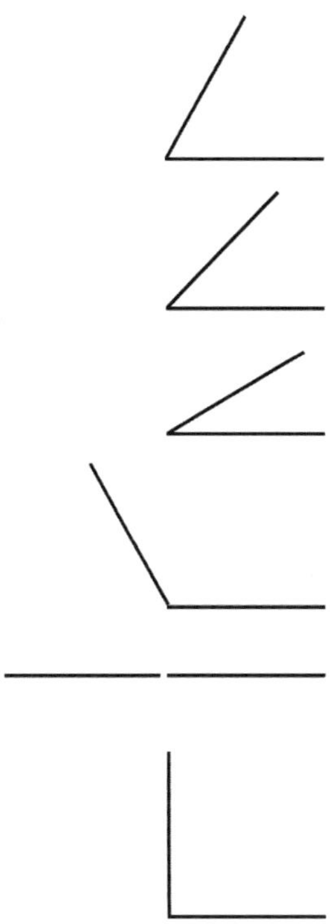

30°

45

60°

90°

120°

180°

135-136. Use a ruler to measure the triangle in tenths of a inch. Then find the perimeter.

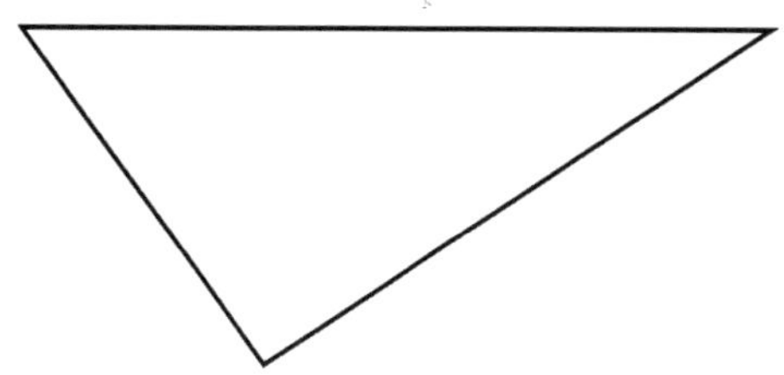

Name: ____________________

Date: ____________________

137-142. Draw lines to match each triangle by sides and angles.

Scalene triangle  Obtuse triangle

Isosceles triangle 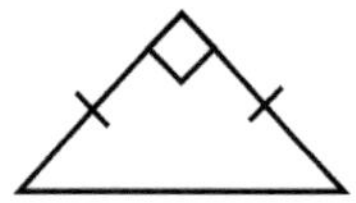 Right triangle

Equilateral triangle 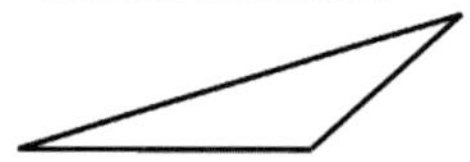 Acute triangle

143-145. Draw all the lines of symmetry in the figures below and answer the questions.

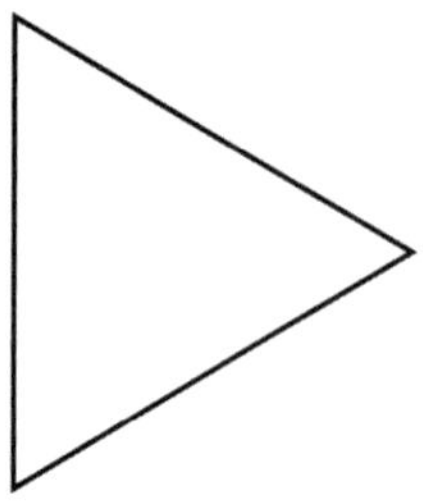

How many lines of symmetry? _____

How many lines of symmetry? _____

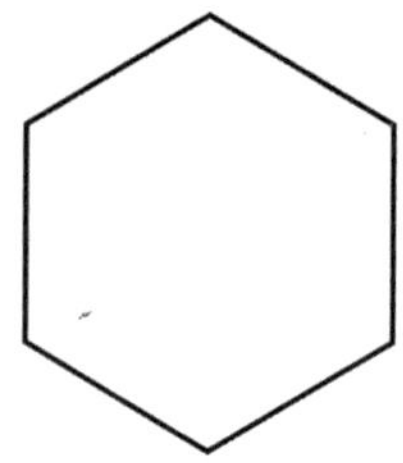

How many lines of symmetry? _____

146-159. Fill in the blanks.

If you turn 360°, where will you end? ____________________

Name the angles in an isosceles right triangle. ____________________

What is special about the sides in an equilateral triangle. ____________________

How many angles does an hexagon have? _________

Can a rectangle also be a square? ___________

Can a parallelogram be a square? ___________

Can a polygon have only two sides? ___________

How many inches are in 2 feet? ________

How many yards is 6 feet? ________

How many centimeters are in 2 decimeters? ________

How many decimeters are in a half a meter? ________

How many centimeters are in a meter? ________

How many milliliters are in a liter? ________

Which is longer, 3 feet or 1 meter? ________

MATH JOURNAL

MATH JOURNAL

MATH JOURNAL

Math Journal

Math Journal

Math Journal

Math Journal

Math Journal

Math Journal

Math Journal

Math Journal

MATH JOURNAL

Math Journal

Math Journal

WARM-UPS

Name: ____________________

Date: ____________________

Warm-Up

Multiply 6 by 2, 3, 4, 5, 6, and 7, then divide by 2, 3, 4, 5, 6, and 7.

					6	
				×	2	
						( )
				×	3	
						( )
				×	4	
						( )
				×	5	
						( )
				×	6	
						( )
				×	7	
						( )
						( )
						( )
						( )
						( )
						( )

Warm-Up

Multiply 7 by 3, 4, 5, 6, 7, and 8, then divide by 3, 4, 5, 6, 7, and 8.

						7	
					×	3	
							( )
					×	4	
							( )
					×	5	
							( )
					×	6	
							( )
					×	7	
							( )
					×	8	
							( )
							( )
							( )
							( )
							( )
							( )

Name: ____________________

Date: ____________________

Warm-Up

Multiply 8 by 3, 4, 5, 6, 7, 8, and 9, then divide by 3, 4, 5, 6, 7, 8, and 9.

8	
× 3	
	( )
× 4	
	( )
× 5	
	( )
× 6	
	( )
× 7	
	( )
× 8	
	( )
× 9	
	( )
	( )
	( )
	( )
	( )
	( )
	( )

Warm-Up

Multiply 9 by 3, 4, 5, 6, 7, 8, and 9, then divide by 3, 4, 5, 6, 7, 8, and 9.

9	
× 3	
	( )
× 4	
	( )
× 5	
	( )
× 6	
	( )
× 7	
	( )
× 8	
	( )
× 9	
	( )
	( )
	( )
	( )
	( )
	( )
	( )

Name: ____________________

Date: ____________________

Warm-Up

Multiply. Use check numbers to check your answers.

58×6 ( ) ( ) ( )

58×40 ( ) ( ) ( )

58×46 ( ) ( ) ( )

279×5 ( ) ( ) ( )

279×30 ( ) ( ) ( )

279×35 ( ) ( ) ( )

Warm-Up

Multiply. Use check numbers to check your answers.

5678×2 ( ) ( ) ( )

5678×70 ( ) ( ) ( )

5678×72 ( ) ( ) ( )

213459×35 ( ) ( ) ( )

Name: ______________________________

Date: ______________________________

Warm-Up

Multiply 11 by 16, 24, 27, and 35, then divide by 2, 3, 4, 5, 6, 7, 8, and 9.

$$
\begin{array}{r}
11 \\
\times\ 16 \\
\hline
\end{array}
$$

$\times\ 24$

$\times\ 27$

$\times\ 35$

Warm-Up

Multiply 12 by 28, 15, 54, and 16, then divide by 2, 3, 4, 5, 6, 7, 8, and 9.

$$
\begin{array}{r}
12 \\
\times\ 28 \\
\hline
\end{array}
$$

$\times\ 15$

$\times\ 54$

$\times\ 16$

Name: ____________________

Date: ____________________

Warm-Up

Multiply 13 by 56, 18, 24, and 15, then divide by 9, 8, 7, 6, 5, 4, 3, and 2.

```
     1 3 (  )
   × 5 6 (  )


         (  )
   × 1 8 (  )


         (  )
   × 2 4 (  )


         (  )
   × 1 5 (  )


         (  )
         (  )
         (  )
         (  )
         (  )
         (  )
         (  )
         (  )
```

Warm-Up

Multiply 14 by 56, 3, 60, and 36, then divide by 9, 8, 7, 6, 5, 4, 3, and 2.

```
     1 4 (  )
   × 5 6 (  )


         (  )
     × 3 (  )
         (  )
   × 6 0 (  )
         (  )
   × 3 6 (  )


         (  )
         (  )
         (  )
         (  )
         (  )
         (  )
         (  )
         (  )
```

Name: ______________________

Date: ______________________

Warm-Up

Multiply 16 by 28, 48, 18, and 15, then divide by 9, 8, 7, 6, 5, 4, 3, and 2.

16 ( )
× 28 ( )

( )
× 48 ( )

( )
× 18 ( )

( )
× 15 ( )

( )
( )
( )
( )
( )
( )
( )
( )

Warm-Up

Multiply 17 by 96, 18, 35, and 6, then divide by 9, 8, 7, 6, 5, 4, 3, and 2.

17 ( )
× 96 ( )

( )
× 18 ( )

( )
× 35 ( )

( )
× 6 ( )
( )
( )
( )
( )
( )
( )
( )
( )

Name: ______________________

Date: ______________________

Warm-Up

Multiply 18 by 35, 27, 16, and 24, then divide by 9, 8, 7, 6, 5, 4, 3, and 2.

```
  1 8 ( )
× 3 5 ( )

      ( )
× 2 7 ( )

      ( )
× 1 6 ( )

      ( )
× 2 4 ( )

      ( )
      ( )
      ( )
      ( )
      ( )
      ( )
      ( )
      ( )
```

Warm-Up

Multiply 19 by 54, 16, 28, and 15, then divide by 9, 8, 7, 6, 5, 4, 3, and 2.

```
  1 9 ( )
× 5 4 ( )

      ( )
× 1 6 ( )

      ( )
× 2 8 ( )

      ( )
× 1 5 ( )

      ( )
      ( )
      ( )
      ( )
      ( )
      ( )
      ( )
      ( )
```

Name: ____________________

Date: ____________________

Warm-Up

Multiply 21 by 24, 18, 15, and 56, then divide by 9, 8, 7, 6, 5, 4, 3, and 2.

```
    2 1 (  )
  × 2 4 (  )


        (  )
  × 1 8 (  )


        (  )
  × 1 5 (  )


        (  )
  × 5 6 (  )


        (  )
        (  )
        (  )
        (  )
        (  )
        (  )
        (  )
        (  )
```

Warm-Up

Multiply 22 by 36, 56, 3, and 60, then divide by 9, 8, 7, 6, 5, 4, 3, and 2.

```
    2 2 (  )
  × 3 6 (  )


        (  )
  × 5 6 (  )


        (  )
    × 3 (  )
        (  )
  × 6 0 (  )
        (  )
        (  )
        (  )
        (  )
        (  )
        (  )
        (  )
        (  )
```

Name: ______________________________

Date: ______________________________

Warm-Up

Multiply 23 by 48, 18, 15, and 28, then divide by 9, 8, 7, 6, 5, 4, 3, and 2.

```
     2 3 (  )
   × 4 8 (  )


         (  )
   × 1 8 (  )


         (  )
   × 1 5 (  )


         (  )
   × 2 8 (  )


         (  )
         (  )
         (  )
         (  )
         (  )
         (  )
         (  )
         (  )
```

Warm-Up

Multiply 24 by 18, 6, 35, and 96, then divide by 9, 8, 7, 6, 5, 4, 3, and 2.

```
     2 4 (  )
   × 1 8 (  )


         (  )
     × 6 (  )
         (  )
   × 3 5 (  )


         (  )
   × 9 6 (  )


         (  )
         (  )
         (  )
         (  )
         (  )
         (  )
         (  )
         (  )
```

Name: ____________________

Date: ____________________

Warm-Up

Multiply 22 by 27, 35, 24, and 16, then divide by 9, 8, 7, 6, 5, 4, 3, and 2.

22	( )
× 27	( )
	( )
× 35	( )
	( )
× 24	( )
	( )
× 16	( )
	( )
	( )
	( )
	( )
	( )
	( )
	( )
	( )

Warm-Up

Multiply 23 by 16, 28, 54, and 15, then divide by 9, 8, 7, 6, 5, 4, 3, and 2.

23	( )
× 16	( )
	( )
× 28	( )
	( )
× 54	( )
	( )
× 15	( )
	( )
	( )
	( )
	( )
	( )
	( )
	( )
	( )

Name: ____________________

Date: ____________________

Warm-Up

Multiply 24 by 15, 24, 56, and 18, then divide by 9, 8, 7, 6, 5, 4, 3, and 2.

```
    2 4 (  )
  × 1 5 (  )


        (  )
  × 2 4 (  )


        (  )
  × 5 6 (  )


        (  )
  × 1 8 (  )


        (  )
        (  )
        (  )
        (  )
        (  )
        (  )
        (  )
        (  )
```

Warm-Up

Multiply 26 by 3, 36, 56, and 60, then divide by 9, 8, 7, 6, 5, 4, 3, and 2.

```
    2 6 (  )
    × 3 (  )
        (  )
  × 3 6 (  )


        (  )
  × 5 6 (  )


        (  )
  × 6 0 (  )
        (  )
        (  )
        (  )
        (  )
        (  )
        (  )
        (  )
        (  )
```

Name: ____________________

Date: ____________________

Warm-Up

Multiply 14 by 96, 90, and 42, then divide by 9, 8, 7, 6, 5, 4, 3, and 2.

$$\begin{array}{r} 14 \; (\;\;) \\ \underline{\times\ 96} \; (\;\;) \end{array}$$

$$\begin{array}{r} (\;\;) \\ \underline{\times\ 90} \; (\;\;) \\ (\;\;) \\ \underline{\times\ 42} \; (\;\;) \end{array}$$

( )
( )
( )
( )
( )
( )
( )
( )

Warm-Up

Multiply 16 by 90, 56, and 72, then divide by 9, 8, 7, 6, 5, 4, 3, and 2.

$$\begin{array}{r} 16 \; (\;\;) \\ \underline{\times\ 90} \; (\;\;) \\ (\;\;) \\ \underline{\times\ 56} \; (\;\;) \end{array}$$

$$\begin{array}{r} (\;\;) \\ \underline{\times\ 72} \; (\;\;) \end{array}$$

( )
( )
( )
( )
( )
( )
( )
( )

Name: ______________________________

Date: ______________________________

Warm-Up

Multiply 17 by 72, 70, and 72 again, then divide by 9, 8, 7, 6, 5, 4, 3, and 2.

17 ( )
× 72 ( )

( )
× 70 ( )
( )
× 72 ( )

( )
( )
( )
( )
( )
( )
( )
( )

Warm-Up

Multiply 18 by 42, 96, and 90, then divide by 9, 8, 7, 6, 5, 4, 3, and 2.

18 ( )
× 42 ( )

( )
× 96 ( )

( )
× 90 ( )
( )
( )
( )
( )
( )
( )
( )
( )

Name: ______________________________

Date: ______________________________

Warm-Up

Multiply 19 by 72, 56, and 90, then divide by 9, 8, 7, 6, 5, 4, 3, and 2.

$$\begin{array}{r} 19 \\ \times\ 72 \\ \hline \end{array}$$

$$\begin{array}{r} \times\ 56 \\ \hline \end{array}$$

$$\begin{array}{r} \times\ 90 \\ \hline \end{array}$$

Warm-Up

Multiply 13 by 72, 72, and 70, then divide by 9, 8, 7, 6, 5, 4, 3, and 2.

$$\begin{array}{r} 13 \\ \times\ 72 \\ \hline \end{array}$$

$$\begin{array}{r} \times\ 72 \\ \hline \end{array}$$

$$\begin{array}{r} \times\ 70 \\ \hline \end{array}$$

Name: ____________________

Date: ____________________

Warm-Up

Multiply 12 by 90, 42, and 96, then divide by 9, 8, 7, 6, 5, 4, 3, and 2.

	1 2 ( )
	× 9 0 ( )
	( )
	× 4 2 ( )
	( )
	× 9 6 ( )
	( )
	( )
	( )
	( )
	( )
	( )
	( )
	( )

Warm-Up

Multiply 11 by 56, 72, and 90, then divide by 9, 8, 7, 6, 5, 4, 3, and 2.

	1 1 ( )
	× 5 6 ( )
	( )
	× 7 2 ( )
	( )
	× 9 0 ( )
	( )
	( )
	( )
	( )
	( )
	( )
	( )
	( )

Certificate of Achievement

Presented to

____

for completing

RIGHTSTART™ MATHEMATICS

LEVEL E

Second Edition

On this __________ *day of* ____________________________

______________________________________*, Teacher*

Joan A. Cotter, Ph.D.

Kathleen Cotter Lawler